10 Florida FAST Grade 6 Math Practice Tests

The Ultimate Test Prep Collection with Answer Explanations

Dr. A. Nazari

10 Practice Tests

The Grand Championship Collection

Welcome, future Math Champion!

*You hold the **ultimate collection** —
ten full-length practice tests designed to take you
from first attempt to **complete mastery**.*

Conquer every Grade 6 topic

Build unshakeable confidence

Rise from Bronze to Gold to Champion

Arrive at test day fully prepared

The championship begins now.

> **Ten tests may seem like
> a marathon, but champi-
> ons are made one step at a
> time. Trust the process!**

 The Champion's Path

Your 4-phase journey from Bronze to Champion

Your warm-up matches. Take these **untimed** to learn the format and set your baseline. Read the answer explanations after each test — this is where you build your foundation.

Silver Round (Tests 4–6)

Set a timer for **75 minutes**. Focus on the topics that tripped you up in Bronze. Practice showing your work on every problem. Your accuracy should be climbing.

Gold Round (Tests 7–9)

Full timed conditions (**60 minutes**). Simulate the real exam environment. Review only the questions you missed — targeted practice is the key to gold.

Championship Final (Test 10)

Your final match. Full exam conditions — timed, quiet, no breaks. This is your victory lap. Show yourself how far you've come!

---

10 Full-Length Practice Tests — every Grade 6 topic

Complete Answer Key with explanations

Formula Reference Sheet

Championship Scoreboard to track your rise

Champion's Tip: Space your tests 2–3 days apart. Use the days in between for targeted review. By Test 10, you'll be amazed at your transformation.

👑 The Champion's Playbook 👑

I   **Read every question twice.** The first read tells you the topic. The second tells you exactly what to solve for. Champions never skim.

II   **Mark the clues.** Circle key numbers, underline the question, and cross out information that's just there to distract you.

III   **Choose your strategy.** Before touching pencil to paper, decide: Am I setting up a ratio? Solving an equation? Finding area? Name the approach.

IV   **Solve, then match.** For multiple choice — work the problem on scratch paper first, then find your answer among the choices.

V   **Eliminate and conquer.** Cross out obviously wrong answers. If you're left with two, you've already doubled your odds. Make an educated pick.

VI   **Estimate to verify.** After solving, ask: "Is this answer reasonable?" A quick mental estimate catches most calculation errors.

VII   **Leave nothing blank.** Even a well-reasoned guess is worth more than an empty space. Use partial work to support your answer.

⏱ Timing Mastery

Tests 1–3: **Untimed** (build foundation)   Tests 4–6: **75 min** (build speed)   Tests 7–10: **60 min** (championship conditions)

⭐ Grade 6 Championship Topics

🏅 Ratios & Proportions   🏅 Integers & Rational Numbers   🏅 Expressions & Equations

🏅 Geometry & Measurement   🏅 Statistics & Data Analysis

> _A true champion isn't someone who never makes mistakes — it's someone who learns from **every single one**. After each test, review your errors carefully. That's where the real growth happens._

Get Online

Find more at
ViewMath.com/FL-Grade6

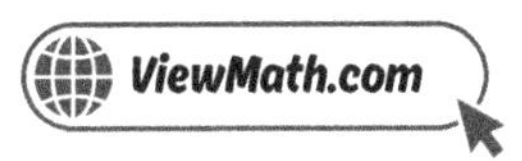
ViewMath.com

♛ The Champion's Toolkit ♛

🏆 Required Equipment

Sharpened Pencils — Two #2 pencils — champions always have a backup

Quality Eraser — A clean, soft eraser that won't smudge your work

Scratch Paper — Blank paper for calculations, diagrams, and number lines

Ruler — Essential for geometry and coordinate plane questions

Timer — Begin using from the Silver Round onward

Quiet Workspace — A calm, well-lit area free from distractions

Permitted in Competition

- ✔ Pencil and eraser
- ✔ Scratch paper (provided)
- ✔ Ruler (if specified)
- ✔ Formula reference in this book

Not Permitted

- ✘ Calculators
- ✘ Electronic devices
- ✘ Textbooks or notes
- ✘ Outside help

- *With 10 tests, space them **2–3 days apart**. This gives time to review mistakes and study between rounds.*

- *Let your child take Tests 1–3 untimed to build familiarity and establish a baseline.*

- *After each test, go through the Answer Key together. Focus on **understanding the reasoning**, not memorizing answers.*

- *Use the Championship Scoreboard to visualize long-term progress. Celebrate improvements at every tier!*

- *Pair with our **Grade 6 Math Study Guide** for topics that need sustained attention.*

Find more at
ViewMath.com/FL-Grade6

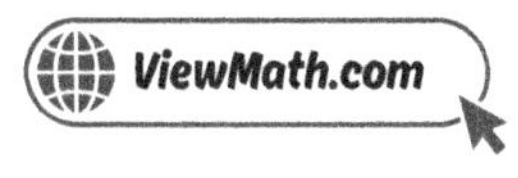

Formula Reference Sheet

◔▲ Area Formulas

Rectangle	$A = l \times w$
Parallelogram	$A = b \times h$
Triangle	$A = \frac{1}{2} \times b \times h$
Trapezoid	$A = \frac{1}{2}(b_1 + b_2) \times h$

⬛ Volume

Rectangular Prism $V = l \times w \times h$

⬛ Surface Area

Find the area of each face, then add them all up.

Rectangular Prism:

$SA = 2lw + 2lh + 2wh$

⬇ Order of Operations

- **P** Parentheses first
- **E** Exponents
- **M/D** Multiply & Divide (left to right)
- **A/S** Add & Subtract (left to right)

% Ratios & Percents

Ratio: $a : b$ or $\frac{a}{b}$

Unit rate: amount per 1 unit

Percent: a ratio out of 100

$Part = Percent \times Whole$

⚖ Integers & Absolute Value

Integers:

$\ldots, -3, -2, -1, 0, 1, 2, 3, \ldots$

$|-5| = 5 \quad |5| = 5$

Absolute value = distance from 0

X¹ Expressions & Equations

Exponent: $3^4 = 3 \times 3 \times 3 \times 3 = 81$

Variable: a letter that stands for a number

Equation: two expressions joined by $=$

Inequality: uses $<, >, \leq, \geq$

⊕ Coordinate Plane

Ordered pair: (x, y)

x-axis: horizontal $\quad$ y-axis: vertical

Origin: $(0, 0)$

Four quadrants (I, II, III, IV)

📊 Statistics

Mean: sum of values $\div$ count

Median: middle value (sorted)

Range: $max - min$

 Championship Scoreboard

Champion's Name: ___________________________________

🏆 Round	🥇 Tier	📅 Date	⭐ Score	😊 Rating
1	Bronze			
2	Bronze			
3	Bronze			
4	Silver			
5	Silver			
6	Silver			
7	Gold			
8	Gold			
9	Gold			
10	👑			

X

My strongest topics (where I consistently score well):

____

____

Topics I improved on the most from Bronze to Gold:

____

____

My score trend (Bronze avg → Gold avg → Championship):

____

____

One strategy that helped me improve the most:

____

____

My confidence level for the real test (1–10): _________ / 10

Find more at
ViewMath.com/FL-Grade6

 Table of Contents

Here's what we'll explore together!

 Let's learn and have fun!

Practice Test 1

☑ 30 Questions

✏ **Before You Start** ✏

✓ **Read each question carefully** before choosing your answer.

✓ **Show your work** on scratch paper when you need to.

✓ **Skip hard questions** and come back to them later.

✓ **Check your answers** when you're done.

✓ **Take your time** — there's no rush!

★ You've Got This! ★

Do your best and show what you know!

1. A garden has flowers and vegetables in a ratio of $4 : 1$. There are 20 flowers. How many vegetables are there?

2. The graph shows the relationship between hours worked and money earned.

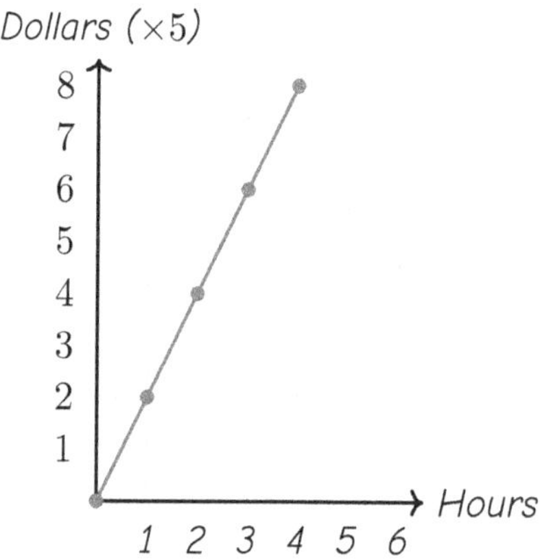

How much money is earned in 6 hours?

A) $30

B) $50

C) $60

D) $40

3. *The table and partial graph below show a ratio relationship.*

x	y
1	4
2	8
3	?

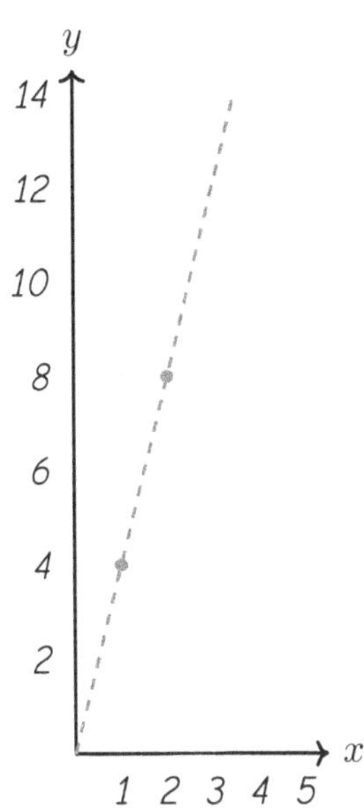

What is the missing value of y when $x = 3$?

(A) 10 (B) 12

(C) 14 (D) 16

4. *What is 72% written as a decimal?*

(A) 72.0 (B) 7.2

(C) 0.072 (D) 0.72

5. The table compares four payment methods.

Method	Uses Your Own Money?	Can Charge Interest?
Cash	Yes	No
Check	Yes	No
Debit Card	Yes	No
Credit Card	No	Yes

Based on the table, which statement is true?

(A) A check borrows money from the bank

(B) A debit card can charge you interest

(C) Only a credit card involves borrowing money

(D) Cash and credit cards work the same way

6. A carpenter has $2\frac{1}{2}$ feet of wood. Each shelf needs $\frac{5}{8}$ of a foot. How many shelves can the carpenter cut?

Your Answer

Find more at
ViewMath.com/FL-Grade6

7. *The bar graph below shows the number of cans a school collected for recycling each week.*

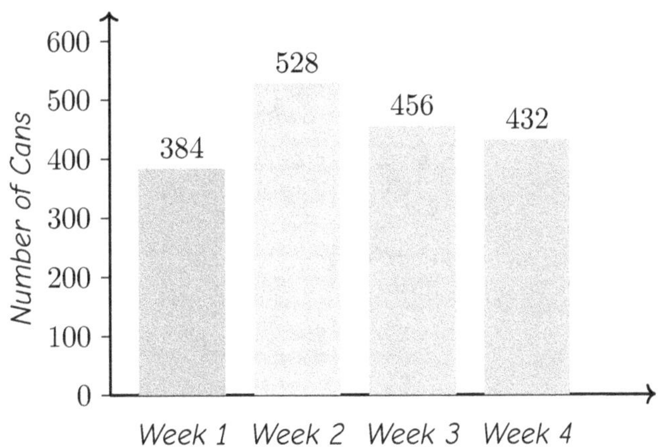

The school wants to divide all the cans equally among 12 classrooms. How many cans does each classroom receive?

(A) 180   (B) 1,800

(C) 150   (D) 130

8. *What is the LCM of 4 and 10?*

(A) 20   (B) 2

(C) 10   (D) 40

9. *Store A's profit last month was −$120 and Store B's profit was −$85. Find each absolute value and state which store was farther from break-even.*

10. *The point $(-7, 3)$ is reflected across the x-axis. Write the coordinates of the reflected point.*

Find more at
ViewMath.com/FL-Grade6

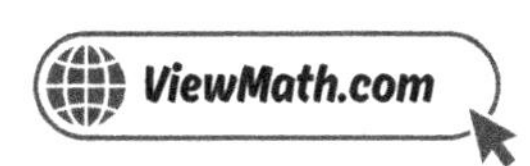

11. *The input-output machine below multiplies the input by -6.*

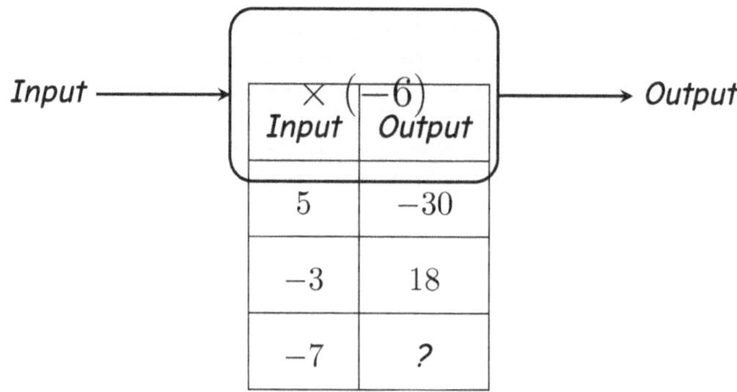

What is the output when the input is -7?

(A) -42

(B) -13

(C) 42

(D) -1

12. *A student is building a cube out of unit cubes. Each edge of the cube is 3 units long. The diagram shows the base layer.*

Base layer (top view)

3 units

3 units

How many unit cubes are needed to build the full cube? Write your answer using an exponent.

Your Answer:

13. *Write a word phrase for the expression $5x + 3$.*

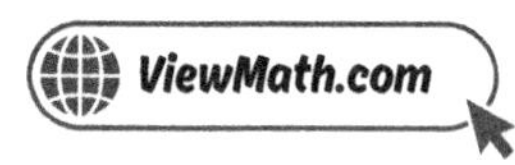

14. How many terms are in the expression $4x + 9 - 2y$?

(A) 2

(B) 3

(C) 4

(D) 5

15. A parking garage charges \$5 plus \$3 per hour. How much does it cost to park for 6 hours? Use the expression $5 + 3h$.

Your Answer

16. Solve: $\dfrac{m}{4} = 7$

(A) $m = 3$

(B) $m = 11$

(C) $m = 28$

(D) $m = 47$

17. Write a real-world situation that can be described by $x \leq 15$.

Your Answer

18. Is $x = 3$ a solution to $x \leq 3$? Is $x = 3$ a solution to $x < 3$? Explain.

Your Answer

19. Which table shows a relationship where y does NOT depend on x?

(A) $x : 1, 2, 3 \quad y : 3, 6, 9$

(B) $x : 1, 2, 3 \quad y : 7, 7, 7$

(C) $x : 1, 2, 3 \quad y : 2, 4, 6$

(D) $x : 1, 2, 3 \quad y : 5, 10, 15$

Find more at
ViewMath.com/FL-Grade6

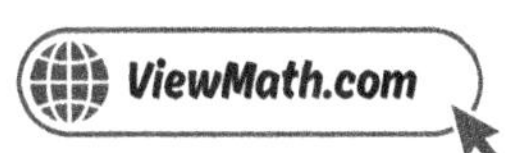

20. A composite figure is made of a rectangle (8 cm by 5 cm) attached to a triangle with base 8 cm and height 3 cm. What is the total area?

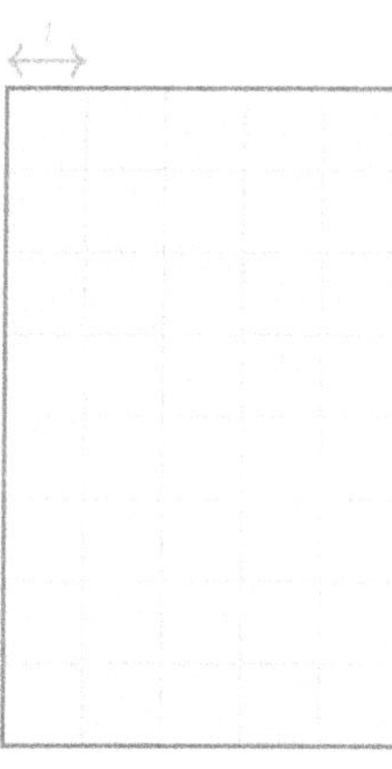

(A) 52 cm^2

(B) 40 cm^2

(C) 55 cm^2

(D) 64 cm^2

21. Volume is measured in which type of units?

(A) Square units (cm^2)

(B) Linear units (cm)

(C) Cubic units (cm^3)

(D) No units are needed

22. Points $A(0, 4)$ and $B(0, -5)$ are connected by a line segment. What is the length of $\overline{AB}$?

(A) 1 unit

(B) 5 units

(C) 4 units

(D) 9 units

23. A rectangular prism has dimensions 10 m, 3 m, and 7 m. What is its surface area?

(A) 210 m^2

(B) 242 m^2

(C) 262 m^2

(D) 121 m^2

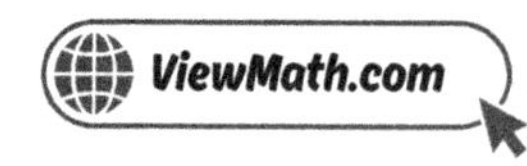

24. If you ask 20 classmates, "How many pets do you have?" and you get the answers $0, 1, 1, 2, 0, 3, 1, 2, 0, 1, 4, 2, 1, 0, 2, 1, 3, 1, 0, 2$, what does this tell you about the question?

(A) It is not a statistical question because some answers are 0.

(B) It is statistical because the answers vary.

(C) It is not statistical because the answers are whole numbers.

(D) It is statistical because there are 20 answers.

25. Data: $5, 10, 15, 20, 25, 30, 35$. What is the range?

(A) 5

(B) 15

(C) 20

(D) 30

26. A dot plot shows hours of TV watched daily: 0 (3 dots), 1 (6 dots), 2 (5 dots), 3 (4 dots), 4 (1 dot), 8 (1 dot).

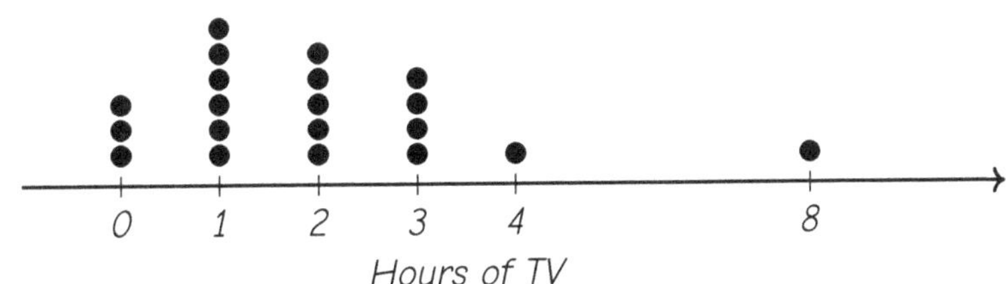

What is the mode, and which value appears to be an outlier?

(A) Mode = 1 hour; outlier = 0

(B) Mode = 1 hour; outlier = 8 hours

(C) Mode = 2 hours; outlier = 8 hours

(D) Mode = 3 hours; outlier = 4 hours

27. Which part of the box plot represents the lowest 25% of the data?

(A) The box

(B) The left whisker (from min to Q1)

(C) The right whisker (from Q3 to max)

(D) The line at the median

Find more at
ViewMath.com/FL-Grade6

28. A student says "The data set with the larger range is always more spread out than the one with the smaller range." Is this always true? Explain.

29. A standard number cube is rolled. What is the probability of rolling a number less than or equal to 3?

(A) $\dfrac{1}{3}$

(B) $\dfrac{3}{6}$

(C) $\dfrac{2}{6}$

(D) $\dfrac{4}{6}$

30. What is the mode of the data in the stem-and-leaf plot below?

Stem	Leaves
3	2 5 5 8
4	1 5 5 5 9
5	0 3

Key: 3 | 2 means 32

Your Answer

Find more at
ViewMath.com/FL-Grade6

ViewMath.com

 End of Practice Test 1 

Great job finishing the test!

☑ My Score

I got _____________ out of 30 questions right.

Check your answers in the Answer Key at the back of the book.

💡 Review any questions you missed. That's how we learn!

📊 Check Your Score Online!

*Visit **ViewMath Academy** to enter your answers and see which topics you need to review. You can also explore lessons, take quizzes, track your scores, and save your progress!*

viewmath.com/score/6.1.FL.16

Or go to viewmath.com/score and enter code: 6.1.FL.16

2

Practice Test 2

☑ 30 Questions

✏ Before You Start ✏

- ✓ **Read each question carefully** before choosing your answer.
- ✓ **Show your work** on scratch paper when you need to.
- ✓ **Skip hard questions** and come back to them later.
- ✓ **Check your answers** when you're done.
- ✓ **Take your time** — there's no rush!

★ You've Got This! ★

Do your best and show what you know!

1. Look at the tape diagram below.

Boys: ▢ ▢ ▢ ▢
Girls: ▢ ▢ ▢ ▢ ▢ ▢

There are 24 girls. How many boys are there?

(A) 4

(B) 12

(C) 16

(D) 20

2. Complete the ratio table below. The ratio of markers to crayons is $4 : 6$.

Markers	Crayons
4	6
8	?
?	24
20	?

Find all three missing values.

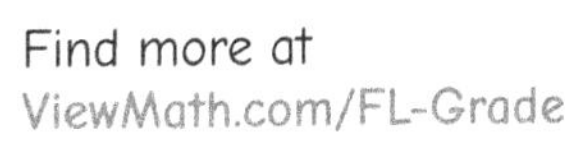

Find more at
ViewMath.com/FL-Grade6

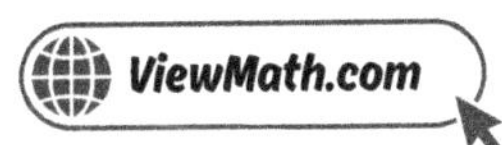

3. Two runners' distances are graphed below.

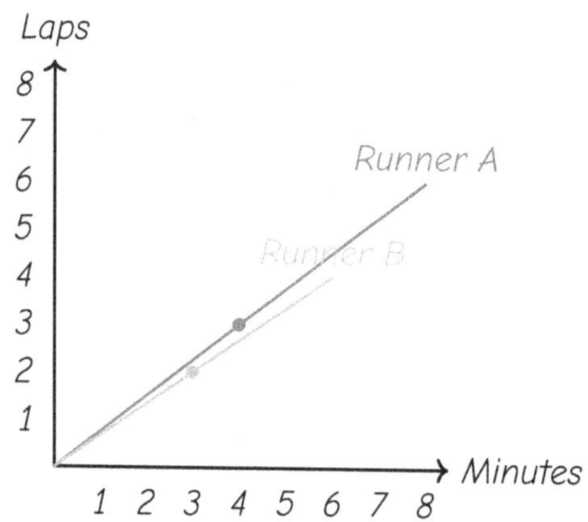

Part A: What is each runner's ratio of laps to minutes?

Part B: Who runs faster? Explain using the graph.

Your Answer:

4. Which of the following is equal to 50%?

(A) $\dfrac{1}{5}$

(B) 0.05

(C) $\dfrac{1}{2}$

(D) 5.0

5. Olivia earns \$240 per month from her part-time job. She saves 20% and spends 10% on school supplies. How much money is left for other expenses?

(A) \$72

(B) \$144

(C) \$168

(D) \$192

Find more at
ViewMath.com/FL-Grade6

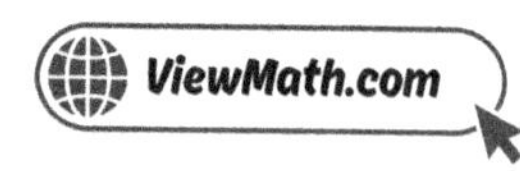

6. Maya jogged $\frac{7}{8}$ of a mile. Each lap around the park is $\frac{1}{4}$ of a mile. How many laps did she complete?

(A) $\frac{7}{32}$

(B) 4

(C) 2

(D) $3\frac{1}{2}$

7. What is $756 \div 3$?

(A) 252

(B) 250

(C) 258

(D) 202

8. Find both the GCF and the LCM of 8 and 12.

9. Which number has the greatest absolute value?

(A) 3

(B) -8

(C) -1

(D) 5

10. Explain why the point $(9, 0)$ is not in any quadrant.

11. A video game player loses 4 points per round for 9 rounds. What is the player's total change in score?

(A) 36

(B) -13

(C) -36

(D) 13

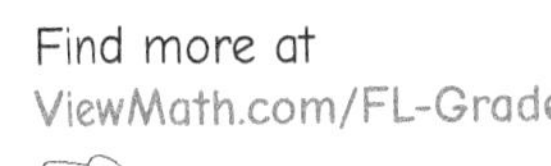

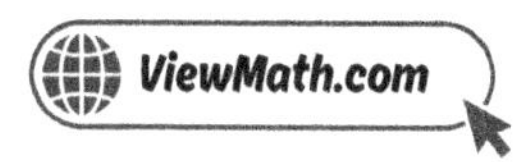

12. *Evaluate:* $(12 - 4)^2 \div 16$

13. *Look at the two-step diagram below. Write the expression that describes the final output when starting with the input n.*

$$\boxed{\text{Input: } n} \longrightarrow \boxed{\text{Multiply by } 3} \longrightarrow \boxed{\text{Subtract } 8} \longrightarrow \boxed{\text{Output: ?}}$$

14. *In the expression $12k + 6$, what are the factors of the term $12k$?*

15. *Evaluate* $\dfrac{x + y}{2}$ *when $x = 10$ and $y = 6$.*

(A) 3

(B) 5

(C) 8

(D) 16

16. *Solve:* $k + 15 = 32$

(A) $k = 47$

(B) $k = 17$

(C) $k = 27$

(D) $k = 2$

17. Which inequality represents "a number y is fewer than 20"?

   (A) $y > 20$ (B) $y \geq 20$

   (C) $y < 20$ (D) $y \leq 20$

18. Which inequality has a graph with an open circle at 0 and shading to the right?

   (A) $x \leq 0$ (B) $x \geq 0$

   (C) $x > 0$ (D) $x < 0$

19. Identify the independent and dependent variables: A runner jogs 6 miles per hour. The distance d depends on how many hours h the runner jogs.

20. A parallelogram has base 12 m and a slant side of 8 m. The height is 6 m. What is the area?

   (A) $96 \ m^2$ (B) $72 \ m^2$

   (C) $48 \ m^2$ (D) $36 \ m^2$

21. A storage container is 5 m long, 3 m wide, and 2 m tall. It is half-full of sand. How many cubic meters of sand are in the container?

   (A) $30 \ m^3$ (B) $15 \ m^3$

   (C) $10 \ m^3$ (D) $60 \ m^3$

22. A rectangle has vertices $(-6, 2)$, $(2, 2)$, $(2, -3)$, $(-6, -3)$. What is the perimeter?

23. How many faces does a rectangular prism have?

(A) 4

(B) 5

(C) 6

(D) 8

24. A student says, "Any question with a number for an answer is statistical." Is the student correct? Explain.

Your Answer:

25. A number line below shows the positions of Q1, the median, and Q3 for a data set.

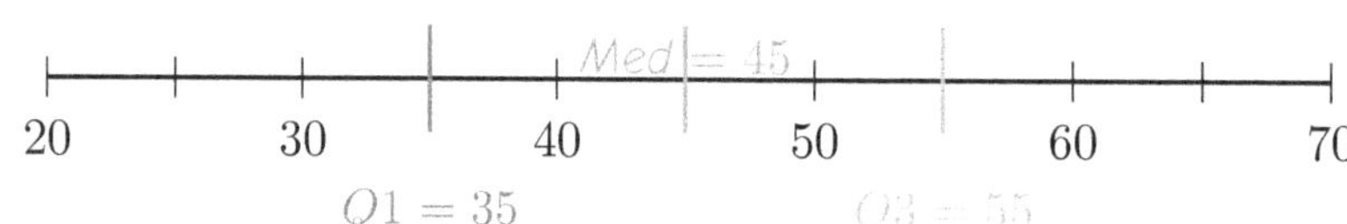

What is the IQR?

(A) 10

(B) 15

(C) 20

(D) 35

26. The frequency table below shows the results of a school survey about how students get to school.

Transportation	Frequency
Bus	45
Car	30
Walk	15
Bike	10

Find the total number of students surveyed, the mode, and the percent who walk to school.

Your Answer:

Find more at
ViewMath.com/FL-Grade6

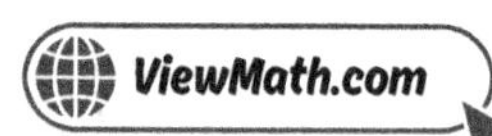

27. *Which five values make up the five-number summary?*

(A) *Mean, median, mode, range, IQR*   (B) *Minimum, Q1, median, Q3, maximum*

(C) *Q1, Q2, Q3, Q4, Q5*   (D) *Mean, Q1, Q3, range, MAD*

28. *A coach records sprint times (seconds) for two athletes over 10 races. Athlete A: median $= 12.5$, range $= 1.2$. Athlete B: median $= 12.5$, range $= 3.0$. Which athlete is more consistent?*

(A) *Athlete A*   (B) *Athlete B*

(C) *They are equally consistent.*   (D) *Cannot be determined without the IQR.*

29. *The probability scale below shows four events. Which event has a probability closest to 0.75?*

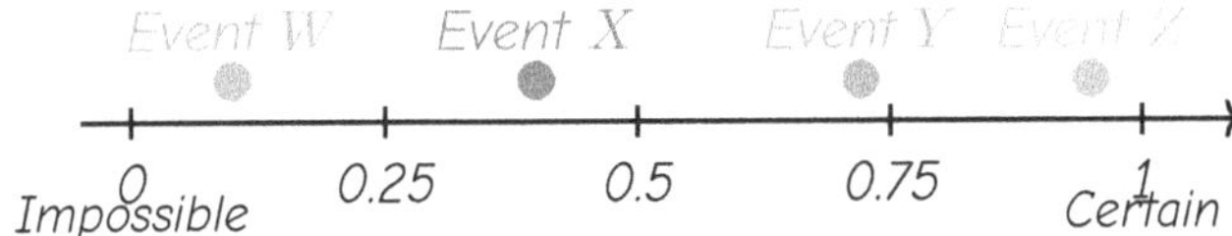

(A) *Event W*   (B) *Event X*

(C) *Event Y*   (D) *Event Z*

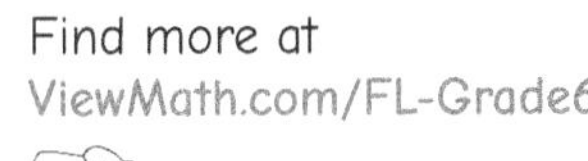
Find more at
ViewMath.com/FL-Grade6

ViewMath.com

30. The back-to-back stem-and-leaf plot below shows the test scores for two classes.

Class A	Stem	Class B
8 5 2	6	3 6
7 4 1	7	0 2 5 9
6 0	8	1 4
	9	2

Key: 2 | 6 | 3 means 62 for Class A and 63 for Class B

Which class has the higher median score?

(A) Class A

(B) Class B

(C) They have the same median.

(D) There is not enough information to tell.

Find more at
ViewMath.com/FL-Grade6

 # End of Practice Test 2 

Great job finishing the test!

 My Score

I got ____________ out of 30 questions right.

Review any questions you missed. That's how we learn!

📊 Check Your Score Online!

Visit **ViewMath Academy** to enter your answers and see which topics you need to review. You can also explore lessons, take quizzes, track your scores, and save your progress!

viewmath.com/score/6.1.FL.17

Or go to viewmath.com/score and enter code: 6.1.FL.17

Practice Test 3

30 Questions

✏️ Before You Start ✏️

- ✓ **Read each question carefully** before choosing your answer.
- ✓ **Show your work** on scratch paper when you need to.
- ✓ **Skip hard questions** and come back to them later.
- ✓ **Check your answers** when you're done.
- ✓ **Take your time** — there's no rush!

⭐ You've Got This! ⭐

Do your best and show what you know!

1. A baker uses 2 eggs **per** 3 cups of flour. Which ratio represents flour to eggs?

(A) $2 : 3$

(B) $3 : 5$

(C) $3 : 2$

(D) $2 : 5$

2. A recipe uses 4 eggs for every 6 cups of flour. How many eggs are needed for 18 cups of flour?

(A) 6

(B) 10

(C) 12

(D) 8

3. A ratio graph passes through $(6, 2)$. What is the value of y when $x = 15$?

(A) 3

(B) 5

(C) 7

(D) 10

4. A store has 200 items. 50% are on sale. How many items are on sale?

(A) 50

(B) 150

(C) 100

(D) 200

5. Emma owes \$500 on a credit card with 18% annual interest. She makes no payments for one year. What is her new balance?

(A) \$518

(B) \$540

(C) \$590

(D) \$680

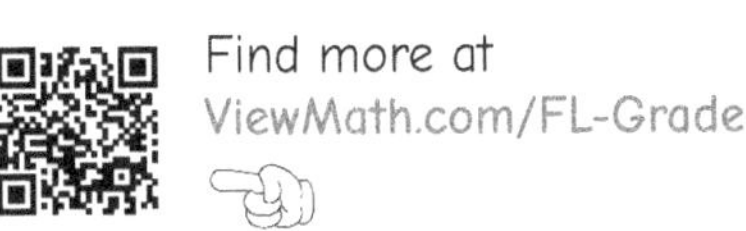

Find more at
ViewMath.com/FL-Grade6

6. What is $\dfrac{3}{5} \div \dfrac{1}{5}$?

   (A) $\dfrac{3}{25}$

   (B) 3

   (C) $\dfrac{1}{3}$

   (D) 15

7. What is $7,344 \div 24$?

   (A) 36

   (B) 360

   (C) 306

   (D) 3,006

8. What is the GCF of 36 and 48?

   (A) 12

   (B) 6

   (C) 24

   (D) 144

9. A company's quarterly result is $-\$200$. A manager says, "We are 200 dollars from break-even." Which mathematical concept did the manager use?

   (A) The opposite of -200

   (B) The absolute value $|-200| = 200$

   (C) Subtracting $200 - 0 = 200$

   (D) The reciprocal of 200

10. Which sign convention describes all points in Quadrant II?

    (A) $(+, +)$

    (B) $(-, +)$

    (C) $(-, -)$

    (D) $(+, -)$

Find more at
ViewMath.com/FL-Grade6

ViewMath.com

11. What is $(-1) \times (-2) \times (-5) \times (-3)$?

A -30 B 30

C -11 D 11

12. Evaluate: $60 \div (3 + 2) - 2^2$

Your Answer

13. Which expression represents "subtract w from 15"?

A $w - 15$ B $15 + w$

C $15 - w$ D $15w$

14. What is the coefficient of w in the expression $9 + w - 3$?

15. Evaluate $2(3y - 1)$ when $y = 5$.

A 28 B 14

C 29 D 10

16. Solve: $x + 24 = 50$

A $x = 26$ B $x = 74$

C $x = 36$ D $x = 2$

17. The table below matches phrases to inequality symbols. Which row has an error?

Row	Phrase	Symbol
1	At least	$\geq$
2	No more than	$\leq$
3	Fewer than	$<$
4	At most	$<$

(A) Row 1

(B) Row 2

(C) Row 3

(D) Row 4

18. The pool opens when the temperature is more than 75°F. Which describes the graph of this inequality?

(A) Closed circle at 75, shade right

(B) Open circle at 75, shade right

(C) Closed circle at 75, shade left

(D) Open circle at 75, shade left

19. A recipe uses 2 cups of flour for every batch of cookies. If you make b batches, which equation gives the total flour f?

(A) $f = b + 2$

(B) $f = 2b$

(C) $b = 2f$

(D) $f = b \div 2$

20. A park is shaped like a trapezoid with parallel sides of 40 m and 60 m and a height of 30 m. What is the area of the park?

(A) $1{,}800\ m^2$

(B) $1{,}200\ m^2$

(C) $2{,}400\ m^2$

(D) $1{,}500\ m^2$

21. A rectangular prism has length $\frac{3}{4}$ ft, width 2 ft, and height 4 ft. What is the volume?

(A) 6 ft^3

(B) 3 ft^3

(C) 24 ft^3

(D) $\frac{3}{2}$ ft^3

22. Find the perimeter of the rectangle shown on the coordinate plane.

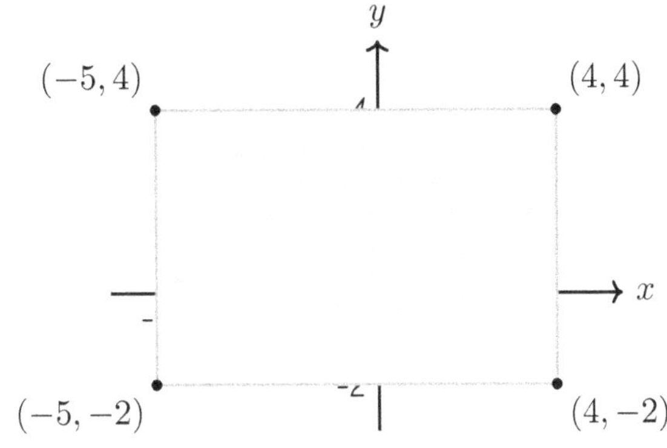

23. A rectangular prism has a surface area of 94 cm^2. Its length is 5 cm, width is 3 cm, and height is 4 cm. Is this correct?

(A) Yes, the surface area is 94 cm^2.

(B) No, the surface area is 120 cm^2.

(C) No, the surface area is 60 cm^2.

(D) No, the surface area is 47 cm^2.

24. Which question would produce data where every answer is the same?

(A) How many hours do you watch TV each day?

(B) How many feet are in one yard?

(C) What is your favorite sport?

(D) How far can you throw a ball?

Find more at
ViewMath.com/FL-Grade6

ViewMath.com

25. *Which statement about MAD is true?*

   (A) *MAD measures the middle value of the data.*

   (B) *MAD tells you the average distance of each value from the mean.*

   (C) *MAD is always larger than the range.*

   (D) *MAD equals the range divided by 2.*

26. *A dot plot of daily steps:* 3000 (1), 4000 (2), 5000 (6), 6000 (4), 7000 (2). *What is the mode? What is the range?*

27. *Data:* $1, 3, 5, 7, 9, 11, 13, 15, 17, 19$. *What is the IQR?*

   (A) 8

   (B) 10

   (C) 14

   (D) 18

28. *A data set has a range of 40 but an IQR of only 8. What does this tell you?*

   (A) *All values are close together.*

   (B) *Most values are clustered, but there are extreme values far from the center.*

   (C) *The median equals the mean.*

   (D) *The data is symmetric.*

29. *A standard number cube (die) is rolled once. What is the probability of rolling a 4?*

   (A) $\dfrac{4}{6}$

   (B) $\dfrac{1}{4}$

   (C) $\dfrac{1}{6}$

   (D) $\dfrac{1}{3}$

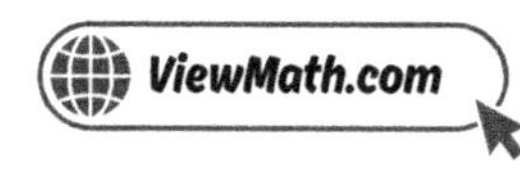

30. A stem-and-leaf plot uses stems $10, 11, 12$ with a key $10 \mid 2 = 102$. What data value does stem 11 with leaf 5 represent?

(A) 15

(B) 115

(C) 150

(D) 1105

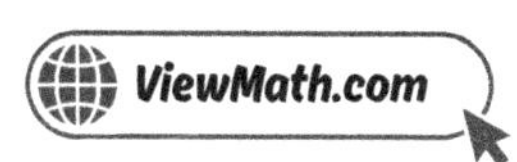

 # End of Practice Test 3 

Great job finishing the test!

 My Score

I got ______________ out of 30 questions right.

Check your answers in the **Answer Key** at the back of the book.

Review any questions you missed. That's how we learn!

📊 Check Your Score Online!

Visit **ViewMath Academy** to enter your answers and see which topics you need to review. You can also explore lessons, take quizzes, track your scores, and save your progress!

viewmath.com/score/6.1.FL.18

Or go to viewmath.com/score and enter code: 6.1.FL.18

Practice Test 4

☑ 30 Questions

✏ Before You Start ✏

✓ **Read each question carefully** before choosing your answer.

✓ **Show your work** on scratch paper when you need to.

✓ **Skip hard questions** and come back to them later.

✓ **Check your answers** when you're done.

✓ **Take your time** — there's no rush!

★ You've Got This! ★

Do your best and show what you know!

1. A store sells 3 pencils **for every** 1 eraser. Which ratio represents pencils to erasers?

   (A) $1:3$ (B) $3:1$

   (C) $3:4$ (D) $1:4$

2. Ella mixes paint in a ratio of 3 parts red to 5 parts white. She uses 24 parts of red. How many total parts of paint does she have?

3. A line passes through the origin and the point $(5, 15)$. What is the ratio x to y?

   (A) $1:5$ (B) $5:1$

   (C) $1:3$ (D) $3:1$

4. Convert 0.875 to a percent and then to a fraction in simplest form.

5. Which of the following is a **need** rather than a **want**?

   (A) A new video game (B) A pair of designer sunglasses

   (C) Groceries for the week (D) Concert tickets

6. What is $\dfrac{2}{3} \div \dfrac{4}{5}$?

   (A) $\dfrac{8}{15}$ (B) $\dfrac{6}{5}$

   (C) $\dfrac{5}{6}$ (D) $\dfrac{2}{15}$

7. In the standard long-division algorithm, the four repeating steps in order are: Divide, _________, Subtract, Bring Down. What is the missing step?

(A) Estimate

(B) Multiply

(C) Check

(D) Add

8. A teacher has 48 pencils and 80 erasers to put into identical prize bags with no supplies left over. What is the greatest number of bags she can make?

9. Mei's account balance is −$18 and Kai's account balance is −$25. Whose balance is closer to $0?

(A) Kai, because $25 > 18$

(B) Mei, because $|-18| = 18 < |-25| = 25$

(C) They are the same distance from $0

(D) Kai, because -25 is farther left on a number line

10. In which quadrant is the point $(-4, 7)$ located?

(A) Quadrant I

(B) Quadrant II

(C) Quadrant III

(D) Quadrant IV

11. A stock loses $3 in value each day for 5 days. Which expression represents the total change in value?

(A) $3 \times 5 = 15$

(B) $(-3) + 5 = 2$

(C) $5 \times (-3) = -15$

(D) $(-5) \times (-3) = 15$

12. Write $9 \times 9 \times 9 \times 9$ using an exponent, then find its value.

Your Answer:

13. The table below shows the relationship between a word phrase and an expression. Which expression belongs in the blank?

Word Phrase	Expression
A number plus 4	$n + 4$
3 times a number	$3n$
8 less than a number	?
A number divided by 2	$n \div 2$

(A) $8 - n$

(B) $n - 8$

(C) $8n$

(D) $n + 8$

14. In the expression $5(y + 8)$, name the two factors.

Your Answer:

15. Evaluate $k^2 + k$ when $k = 3$.

(A) 6

(B) 9

(C) 12

(D) 15

16. The number line below shows a jump. Which equation does this model represent?

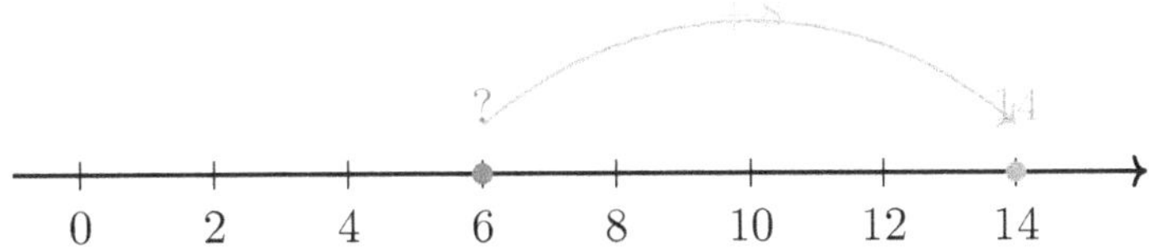

(A) $x + 8 = 14$, so $x = 6$

(B) $x - 8 = 14$, so $x = 22$

(C) $8x = 14$, so $x = 1.75$

(D) $x + 14 = 8$, so $x = -6$

17. Which of the following values is NOT a solution to $y \geq 3$?

(A) $y = 3$

(B) $y = 4$

(C) $y = 100$

(D) $y = 2.5$

18. Describe the graph of $n \leq 5$.

(A) Open circle at 5, shade left

(B) Closed circle at 5, shade left

(C) Open circle at 5, shade right

(D) Closed circle at 5, shade right

19. The number of inches i equals 12 times the number of feet f: $i = 12f$. How many inches are in 5 feet?

(A) 17 inches

(B) 48 inches

(C) 60 inches

(D) 125 inches

20. In a parallelogram, which measurement is used as the height?

(A) The slanted side

(B) The diagonal

(C) The perpendicular distance between the base and its opposite side

(D) The longest side

21. A rectangular prism has a volume of 120 cm^3. Its length is 10 cm and width is 4 cm. What is the height?

- (A) 3 cm
- (B) 12 cm
- (C) 6 cm
- (D) 30 cm

22. A rectangle has vertices $(-5, 4)$, $(3, 4)$, $(3, -2)$, and $(-5, -2)$. What is the perimeter?

- (A) 14 units
- (B) 22 units
- (C) 28 units
- (D) 48 units

23. Which net below could fold into a cube?

A   B   C

- (A) Net A only
- (B) Net B only
- (C) Net C only
- (D) Nets A and B

24. Explain why "What is my teacher's age?" is **not** a statistical question.

Your Answer:

25. Data (in order): $4, 6, 8, 10, 12, 14, 16$. What is Q3?

- (A) 10
- (B) 12
- (C) 14
- (D) 16

26. A dot plot of temperatures shows: 68°F (1 dot), 70°F (3 dots), 72°F (4 dots), 74°F (2 dots). What is the range?

(A) 4

(B) 6

(C) 10

(D) 74

27. If the whiskers of a box plot are both short and the box is narrow, what does this tell you?

(A) The data is very spread out.

(B) The data is highly consistent with little variability.

(C) There are many outliers.

(D) The data set is very large.

28. Data: $12, 15, 18, 20, 22, 25, 28, 30$. What is the IQR?

(A) 10

(B) 11

(C) 18

(D) 8

29. Use the probability scale below. Place each event at the correct position by writing its letter on the scale. Then state each probability as a fraction or decimal.

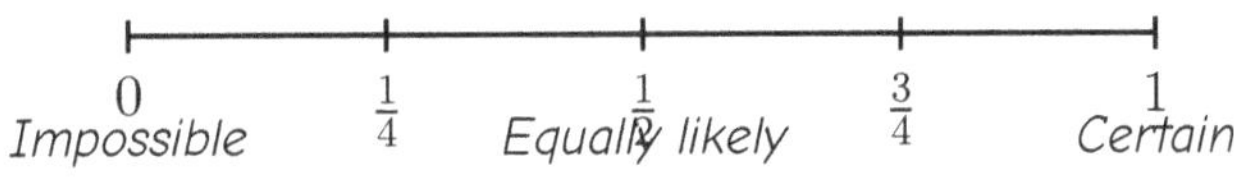

Event A: Rolling a 7 on a standard number cube.

Event B: Flipping tails on a fair coin.

Event C: Drawing a red marble from a bag that has 6 red and 2 blue marbles.

Event D: Rolling a number less than 7 on a standard number cube.

30. *The data set is:* $32, 35, 41, 44, 48, 53, 57.$ *Which stem in the stem-and-leaf plot would have the most leaves?*

(A) 3

(B) 4

(C) 5

(D) *All stems have the same number of leaves.*

 # End of Practice Test 4 

Great job finishing the test!

 My Score

I got _____________ out of 30 questions right.

Check your answers in the Answer Key at the back of the book.

Review any questions you missed. That's how we learn!

📊 Check Your Score Online!

Visit **ViewMath Academy** to enter your answers and see which topics you need to review. You can also explore lessons, take quizzes, track your scores, and save your progress!

viewmath.com/score/6.1.FL.19

Or go to viewmath.com/score and enter code: 6.1.FL.19

5

Practice Test 5

30 Questions

✎ **Before You Start** ✎

✔ **Read each question carefully** before choosing your answer.

✔ **Show your work** on scratch paper when you need to.

✔ **Skip hard questions** and come back to them later.

✔ **Check your answers** when you're done.

✔ **Take your time** — there's no rush!

★ You've Got This! ★

Do your best and show what you know!

1. A class votes on two activities: hiking and swimming. The ratio of votes for hiking to swimming is $3 : 2$. There are 25 votes total. How many voted for swimming?

2. Which pair of ratios are equivalent?

   (A) $2 : 3$ and $4 : 9$

   (B) $2 : 3$ and $6 : 9$

   (C) $2 : 3$ and $8 : 9$

   (D) $2 : 3$ and $3 : 2$

3. A graph of equivalent ratios passes through $(0, 0)$ and $(3, 2)$. Which statement is true?

   (A) The point $(9, 4)$ is on the line.

   (B) The point $(6, 4)$ is on the line.

   (C) The point $(6, 5)$ is on the line.

   (D) The point $(9, 8)$ is on the line.

4. The bar model below shows the results of a class vote on a field trip destination.

Zoo (60%)	Farm	Park

   Total: 100%

   If the Farm section is twice as wide as the Park section, what percent voted for the Farm?

   (A) 10%

   (B) 40%

   (C) $\dfrac{20}{3}\%$

   (D) $\approx 27\%$

5. Which payment method lets you borrow money that you must pay back later, possibly with interest?

   (A) Cash

   (B) Check

   (C) Credit card

   (D) Debit card

Find more at
ViewMath.com/FL-Grade6

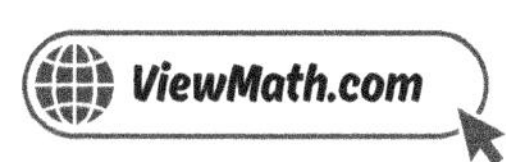

6. A board is $\frac{7}{8}$ of a meter long. You need pieces that are $\frac{1}{4}$ of a meter. How many pieces can you cut? Express your answer as a fraction or mixed number.

Your Answer:

7. What is $1{,}575 \div 5$?

(A) 305

(B) 3,150

(C) 315

(D) 351

8. You have a ribbon that is 72 inches long and another that is 96 inches long. You want to cut both ribbons into equal-length pieces with no ribbon left over. What is the longest possible piece?

(A) 24

(B) 8

(C) 288

(D) 12

9. Amy owes \$15 and Beth owes \$22. Their debts are written as −\$15 and −\$22. Which expression correctly shows that Beth owes more?

(A) $-22 > -15$

(B) $|-22| > |-15|$

(C) $-15 > -22$, so Amy owes more

(D) $|-15| > |-22|$

10. Starting at the origin, you move 4 units to the left and 3 units up, then 7 units to the right and 5 units down. What point do you end at?

Your Answer:

Find more at
ViewMath.com/FL-Grade6

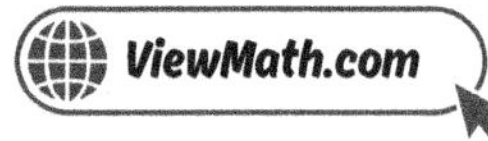

11. *What is $(-9) \times 7$?*

12. *Evaluate: $3 + 5 \times 2$*

   (A) 16   (B) 13

   (C) 11   (D) 30

13. *Write an expression for: "a number t divided by 4, then subtract 1."*

14. *Which expression has exactly 4 terms?*

   (A) $2x + 3y$ (B) $a + b + c$

   (C) $5m - 2n + 7 + m$ (D) $4p$

15. *Evaluate $20 - 3n$ when $n = 4$.*

   (A) 8   (B) 12

   (C) 17   (D) 32

16. *Solve: $\dfrac{w}{8} = 9$*

Find more at
ViewMath.com/FL-Grade6

ViewMath.com

17. Which phrase matches the inequality $t \le 30$?

   (A) The temperature is more than 30 degrees   (B) The temperature is exactly 30 degrees

   (C) The temperature is no more than 30 degrees   (D) The temperature is at least 30 degrees

18. Which inequality matches a graph with a closed circle at 3 and shading to the right?

   (A) $x > 3$ (B) $x < 3$

   (C) $x \ge 3$ (D) $x \le 3$

19. In the equation $y = x + 4$, which statement is true?

   (A) x is the dependent variable   (B) y is always 4

   (C) x is the independent variable   (D) y decreases as x increases

20. A trapezoid has bases $b_1 = 4$ cm and $b_2 = 4$ cm with height 6 cm. What shape is this trapezoid actually equivalent to?

   (A) A triangle   (B) A parallelogram

   (C) A circle   (D) A pentagon

21. A box is 2.5 cm long, 4 cm wide, and 6 cm tall. What is the volume?

   Your Answer:

22. A rectangle has vertices $(1, 2)$, $(7, 2)$, $(7, -4)$, and $(1, -4)$. What is the length of the longer side?

   (A) 5 units   (B) 6 units

   (C) 7 units   (D) 8 units

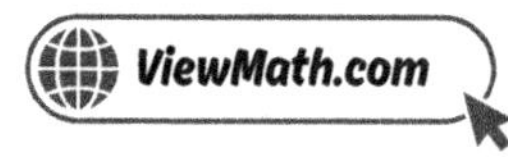

23. *A box is 5 in long, 5 in wide, and 10 in tall. What is the surface area?*

(A) $250\ in^2$

(B) $200\ in^2$

(C) $300\ in^2$

(D) $150\ in^2$

24. *Which of the following is a statistical question?*

(A) How many days are in February during a leap year?

(B) How many hours do sixth graders spend on homework each week?

(C) What is the boiling point of water in degrees Celsius?

(D) How many states are in the United States?

25. *Data:* $10, 10, 10, 10, 10$. *What is the MAD?*

(A) 0

(B) 2

(C) 5

(D) 10

26. *A data set has two values that each appear 4 times, and no other value appears more than 3 times. How many modes does this data set have?*

(A) 0

(B) 1

(C) 2

(D) 4

27. *A box plot shows: min* $= 25$, $Q1 = 35$, *median* $= 50$, $Q3 = 65$, *max* $= 80$. *Find the range and IQR.*

Your Answer

28. *Store A daily sales: median = $500, range = $200. Store B daily sales: median = $500, range = $800. What can you conclude?*

(A) Both stores are equally consistent.

(B) Store A is more consistent in its daily sales.

(C) Store B is more consistent in its daily sales.

(D) Store B has lower typical sales.

29. *The spinner below has 8 equal sections. Find each probability as a fraction in simplest form.*
(a) P(red)   (b) P(blue)   (c) P(not green)

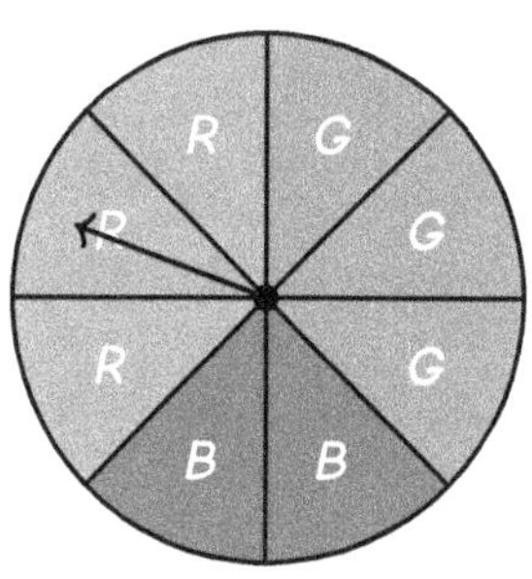

30. *A student made the stem-and-leaf plot below. What is wrong with it?*

Stem	Leaves
5	3 1 7 9
6	2 4

Key: 5 | 3 *means* 53

(A) The stems are out of order.

(B) The leaves for stem 5 are not in order from least to greatest.

(C) The key is incorrect.

(D) There are too few stems.

 End of Practice Test 5 

Great job finishing the test!

✓ My Score

I got ____________ out of 30 questions right.

Check your answers in the Answer Key at the back of the book.

Review any questions you missed. That's how we learn!

📊 Check Your Score Online!

Visit **ViewMath Academy** to enter your answers and see which topics you need to review. You can also explore lessons, take quizzes, track your scores, and save your progress!

viewmath.com/score/6.1.FL.20

Or go to viewmath.com/score and enter code: 6.1.FL.20

Practice Test 6

 30 Questions

✏ Before You Start ✏

✓ **Read each question carefully** before choosing your answer.

✓ **Show your work** on scratch paper when you need to.

✓ **Skip hard questions** and come back to them later.

✓ **Check your answers** when you're done.

✓ **Take your time** — there's no rush!

 You've Got This! 

Do your best and show what you know!

1. A smoothie recipe uses 3 bananas **for every** 4 cups of yogurt. Write the ratio of bananas to yogurt. Then write the ratio of yogurt to bananas.

2. Look at this ratio table. What is the missing value?

Apples	Oranges
3	5
6	?

A) 8   B) 10

C) 15   D) 11

3. Which table matches a graph that passes through $(0,0)$, $(2,3)$, and $(4,6)$?

A)

x	y
2	3
6	9

B)

x	y
2	3
6	8

C)

x	y
3	2
6	9

D)

x	y
2	4
6	12

4. What is 0.35 written as a percent?

A) 0.35%   B) 3.5%

C) 35%   D) 350%

5. Which is the best reason to save money regularly?

   (A) So you can spend more on wants right away

   (B) So you have money for unexpected expenses or future goals

   (C) So the bank will give you a credit card

   (D) So you never have to earn income again

6. What is the reciprocal of $\frac{3}{7}$?

   (A) $\frac{3}{7}$

   (B) $\frac{7}{3}$

   (C) $-\frac{3}{7}$

   (D) $\frac{1}{3}$

7. A baker makes 2,016 cookies and packs 14 cookies per bag. How many bags can she fill?

   (A) 142

   (B) 144

   (C) 148

   (D) 104

8. What is the LCM of 3 and 9?

   (A) 9

   (B) 3

   (C) 27

   (D) 18

9. What is the opposite of -4?

   (A) -4

   (B) 0

   (C) $\frac{1}{4}$

   (D) 4

Find more at
ViewMath.com/FL-Grade6

ViewMath.com

10. *Starting at the origin, how do you plot the point $(-3, 4)$?*

(A) *Move 3 units right and 4 units up*

(B) *Move 3 units left and 4 units down*

(C) *Move 3 units left and 4 units up*

(D) *Move 4 units left and 3 units up*

11. *What is $(-2) \times 5 \times (-3) \div (-6)$?*

(A) 5

(B) -5

(C) -30

(D) 30

12. *A student evaluated $2 + 5 \times 3^2$ step by step. Look at the work shown below. In which step did the student make a mistake?*

Step 1: $\quad 2 + 5 \times 3^2$

Step 2: $\quad 7 \times 3^2$

Step 3: $\quad 7 \times 9$

Step 4: $\quad 63$

(A) *Step 1*

(B) *Step 2*

(C) *Step 3*

(D) *Step 4*

13. *Write an expression for: "the sum of a and b, divided by 2."*

14. *In $3(p + 4)$, the group $(p + 4)$ acts as which part of the expression?*

(A) *A term*

(B) *A coefficient*

(C) *A constant*

(D) *A factor*

Find more at
ViewMath.com/FL-Grade6

ViewMath.com

15. Evaluate $\dfrac{n}{3} + 8$ when $n = 15$.

    (A) 5   (B) 11

    (C) 13   (D) 23

16. Solve: $x + 3.5 = 10$

    (A) $x = 6.5$ (B) $x = 13.5$

    (C) $x = 7.5$ (D) $x = 3.5$

17. Write an inequality: A suitcase must weigh less than 50 pounds. Let w = weight.

18. Pick two values that ARE solutions to $x > -1$ and one value that is NOT.

Your Answer

19. A car wash charges \$10 per car. Which variable goes on the x-axis when graphing?

    (A) Total cost   (B) Number of cars

    (C) Price per car   (D) Profit

Find more at
ViewMath.com/FL-Grade6

ViewMath.com

20. A kite-shaped sign can be split into two triangles, each with base 6 in and height 8 in. What is the total area of the sign?

21. If you double the height of a rectangular prism but keep the length and width the same, what happens to the volume?

(A) The volume stays the same.

(B) The volume doubles.

(C) The volume triples.

(D) The volume quadruples.

22. Points $A(-1, -4)$ and $B(-1, 8)$ form a vertical segment. What is its length?

23. A rectangular prism has length 8 cm, width 5 cm, and height 2 cm. Leo says the surface area is 80 cm^2. What did Leo calculate instead?

(A) The volume

(B) The perimeter

(C) Half the surface area

(D) The area of just one face

24. *Which of the following is a statistical question?*

    (A) *What year was the school built?*   (B) *What is the name of our principal?*

    (C) *How many minutes does each student exercise per day?*   (D) *How many continents are there?*

25. *Data: $20, 22, 24, 26, 28$. The mean is 24. What is the MAD?*

    (A) 2   (B) 2.4

    (C) 4   (D) 8

26. *Which display is best for showing favorite colors of students?*

    (A) *Histogram*   (B) *Dot plot*

    (C) *Frequency table or bar graph*   (D) *Box plot*

27. *Which data set has the five-number summary: $5, 10, 15, 20, 25$?*

    (A) $5, 10, 15, 20, 25$ (B) $5, 8, 10, 15, 18, 20, 22, 25$

    (C) $5, 7, 10, 12, 15, 18, 20, 23, 25$ (D) *Any of these could have that five-number summary.*

28. *Daily steps for Week 1: median $= 8,000$, IQR $= 2,000$. Week 2: median $= 9,500$, IQR $= 1,200$. Which week shows a higher typical step count and more consistent activity?*

    (A) *Week 1 for both*   (B) *Week 2 for both*

    (C) *Week 1 higher, Week 2 more consistent*   (D) *Week 2 higher, Week 1 more consistent*

Find more at
ViewMath.com/FL-Grade6

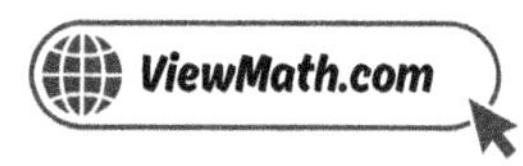

29. *A fair coin is flipped once. What is the probability of landing on heads?*

(A) $\dfrac{1}{4}$

(B) $\dfrac{1}{3}$

(C) $\dfrac{1}{2}$

(D) 1

30. *In a stem-and-leaf plot, the stem represents —*

(A) *the last digit of each data value*

(B) *all digits except the last digit of each data value*

(C) *the average of the data set*

(D) *the number of data values*

Find more at
ViewMath.com/FL-Grade6

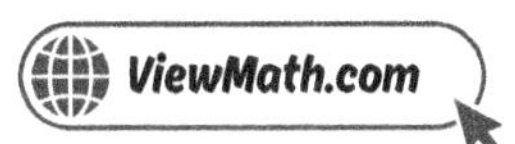

 End of Practice Test 6 

Great job finishing the test!

 My Score

I got _____________ out of 30 questions right.

Check your answers in the Answer Key at the back of the book.

Review any questions you missed. That's how we learn!

 Check Your Score Online!

Visit **ViewMath Academy** to enter your answers and see which topics you need to review. You can also explore lessons, take quizzes, track your scores, and save your progress!

viewmath.com/score/6.1.FL.21

Or go to *viewmath.com/score* and enter code: 6.1.FL.21

Practice Test 7

☑ 30 Questions

✏️ Before You Start ✏️

- ✔ **Read each question carefully** before choosing your answer.
- ✔ **Show your work** on scratch paper when you need to.
- ✔ **Skip hard questions** and come back to them later.
- ✔ **Check your answers** when you're done.
- ✔ **Take your time** — there's no rush!

⭐ You've Got This! ⭐

Do your best and show what you know!

1. A lemonade stand sells 7 cups of lemonade **for each** 2 cups of iced tea. If they sold 21 cups of lemonade, how many cups of iced tea did they sell?

2. A ratio table for pens to notebooks starts with $2 : 3$. Which of these is a valid row?

   (A) $4 : 5$ (B) $8 : 12$

   (C) $6 : 12$ (D) $10 : 12$

3. A ratio graph passes through $(4, 10)$ and the origin. List two other points on this line.

4. A class has 30 students. 70% passed a quiz. How many students passed?

   (A) 7   (B) 21

   (C) 23   (D) 70

5. Explain one advantage and one disadvantage of using a credit card instead of cash.

6. What is $\dfrac{1}{2} \div \dfrac{1}{4}$?

   (A) $\dfrac{1}{8}$ (B) 8

   (C) 2   (D) $\dfrac{1}{2}$

Find more at
ViewMath.com/FL-Grade6

ViewMath.com

7. Compute $9,072 \div 42$.

8. What is the GCF of 60 and 84?

(A) 4

(B) 6

(C) 12

(D) 420

9. What is the opposite of 9?

(A) 9

(B) -9

(C) 0

(D) $\dfrac{1}{9}$

10. Where is the point $(-6, 0)$ located on the coordinate plane?

(A) On the y-axis

(B) On the x-axis

(C) In Quadrant II

(D) In Quadrant III

11. A company's stock drops \$8 in value each day for 5 days. What is the total change in the stock's value?

12. Which shows $5 \times 5 \times 5$ written using an exponent?

(A) $5 + 3$

(B) 5×3

(C) 3^5

(D) 5^3

Find more at
ViewMath.com/FL-Grade6

13. A movie ticket costs d dollars. Which expression represents the cost of 5 tickets?

(A) $d + 5$

(B) $d - 5$

(C) $5d$

(D) $d \div 5$

14. How many constants are in $5x + 3 + 2y + 8$?

(A) 1

(B) 2

(C) 3

(D) 4

15. Evaluate $8y - 3$ when $y = 6$.

Your Answer:

16. Solve: $5n = 45$

(A) $n = 9$

(B) $n = 40$

(C) $n = 50$

(D) $n = 225$

17. Which inequality represents "a number x is greater than 7"?

(A) $x < 7$

(B) $x > 7$

(C) $x \leq 7$

(D) $x = 7$

18. A student graphs $x > 2$ with a closed circle at 2 and shading to the right. What is the error?

(A) Should shade to the left

(B) Should use an open circle at 2

(C) Should use a circle at 0 instead

(D) There is no error

Find more at
ViewMath.com/FL-Grade6

19. A phone battery loses 5% each hour. Which is the independent variable?

(A) Battery percentage

(B) Phone model

(C) Number of hours

(D) Screen brightness

20. A trapezoid has an area of 84 cm^2 and bases of 10 cm and 14 cm. What is the height?

21. A toy box is 3 ft long, 2 ft wide, and 2 ft tall. What is the volume?

(A) $7\ ft^3$

(B) $12\ ft^3$

(C) $6\ ft^3$

(D) $24\ ft^3$

22. To find the length of a horizontal side on the coordinate plane, you subtract the:

(A) y-coordinates and take the absolute value

(B) x-coordinates and take the absolute value

(C) x-coordinate from the y-coordinate

(D) coordinates and divide by 2

23. Find the surface area of a rectangular prism with length 7 m, width 5 m, and height 3 m.

24. *A student surveyed classmates about how many books they read last month and made the dot plot below.*

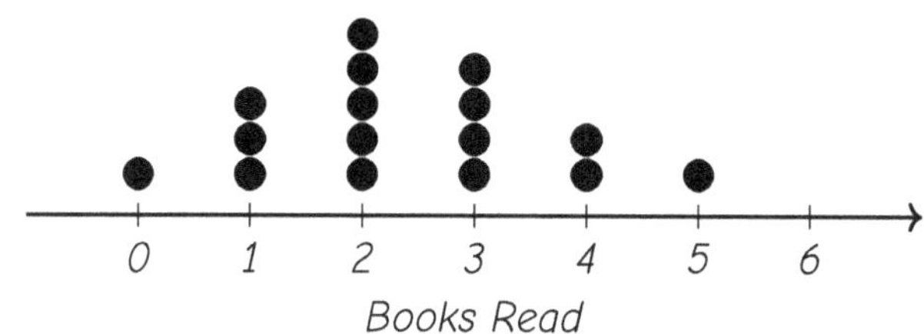

What does this dot plot confirm about the survey question "How many books did you read last month?"

(A) *It is not statistical because only whole numbers appear.*

(B) *It is statistical because the answers vary from 0 to 5.*

(C) *It is not statistical because some students read the same number.*

(D) *It is statistical because there are exactly 16 dots.*

25. *Data:* $100, 105, 110, 115, 120, 125, 130.$ *Find the IQR.*

26. *A data set is:* $5, 5, 5, 5, 5.$ *What is the mode?*

(A) 0

(B) 1

(C) 5

(D) *There is no mode.*

27. The two box plots below represent daily sales (in dollars) at two stores over the same month.

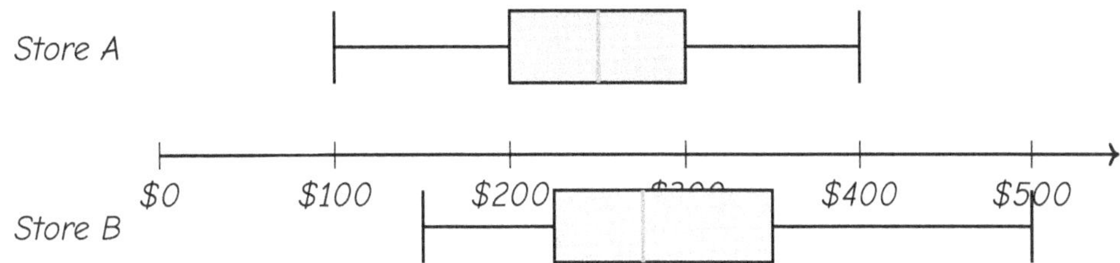

Compare the two stores. Which store has higher typical sales? Which has more predictable sales? Use the box plots to support your answers.

Your Answer

28. Which two measures together best describe the "center" and "spread" of a data set?

(A) Mean and mode

(B) Median and IQR

(C) Range and maximum

(D) Minimum and maximum

29. Event A has a probability of $\frac{2}{5}$ and Event B has a probability of $\frac{3}{4}$. Which statement is true?

(A) Event A is more likely than Event B

(B) Event B is more likely than Event A

(C) Both events are equally likely

(D) Neither event can happen

Find more at
ViewMath.com/FL-Grade6

30. In the plot below, one leaf marked with "?" is missing. The median of the full data set is 37. What is the missing leaf?

Stem	Leaves
2	0 4
3	1 5 ?
4	2 6 8
5	3

Key: 2 | 0 means 20

(A) 5   (B) 6

(C) 7   (D) 9

 # End of Practice Test 7 

Great job finishing the test!

 My Score

I got __________ out of 30 questions right.

Check your answers in the Answer Key at the back of the book.

💡 *Review any questions you missed. That's how we learn!*

📊 Check Your Score Online!

*Visit **ViewMath Academy** to enter your answers and see which topics you need to review. You can also explore lessons, take quizzes, track your scores, and save your progress!*

viewmath.com/score/6.1.FL.22

Or go to viewmath.com/score and enter code: 6.1.FL.22

Practice Test 8

30 Questions

✏️ Before You Start ✏️

- ✔ **Read each question carefully** before choosing your answer.
- ✔ **Show your work** on scratch paper when you need to.
- ✔ **Skip hard questions** and come back to them later.
- ✔ **Check your answers** when you're done.
- ✔ **Take your time** — there's no rush!

⭐ You've Got This! ⭐

Do your best and show what you know!

1. The ratio of cats to dogs at a pet store is 3 : 4. Each part in the tape diagram represents 2 animals. How many dogs are there?

(A) 6

(B) 8

(C) 3

(D) 4

2. The graph below shows points that represent equivalent ratios of flour to sugar in a recipe.

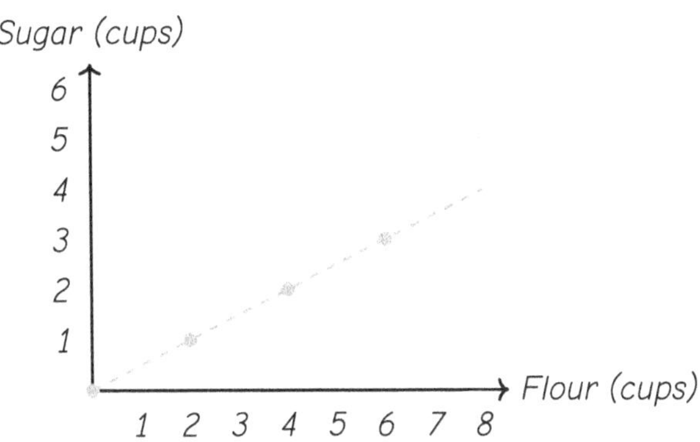

Part A: What is the ratio of flour to sugar?

Part B: If you need 8 cups of flour, how many cups of sugar do you need?

3. True or false: a graph of equivalent ratios can be a curved line. Explain.

4. Which of the following is 100% as a fraction?

(A) $\dfrac{1}{100}$

(B) $\dfrac{100}{10}$

(C) $\dfrac{100}{100}$

(D) $\dfrac{10}{100}$

Find more at
ViewMath.com/FL-Grade6

ViewMath.com

5. Nina puts $800 in a savings account that earns 6% simple interest per year. How much interest does she earn after 2 years, and what is the total amount in her account?

6. A student solved the problem below using "Keep, Change, Flip." Look at the student's work.

$$\textbf{Problem:} \quad \frac{4}{5} \div \frac{2}{3}$$

$$\textbf{Step 1:} \quad \text{Keep } \frac{4}{5} \quad \text{Change } \div \text{ to } \times$$

$$\textbf{Step 2:} \quad \text{Flip } \frac{4}{5} \text{ to get } \frac{5}{4}$$

$$\textbf{Step 3:} \quad \frac{5}{4} \times \frac{2}{3} = \frac{10}{12} = \frac{5}{6}$$

What error did the student make?

(A) The student forgot to change $\div$ to $\times$

(B) The student made a multiplication error in Step 3

(C) The student forgot to simplify

(D) The student flipped the wrong fraction

7. Which multiplication equation can be used to check that $4{,}752 \div 12 = 396$?

(A) $396 \times 12 = 4{,}752$

(B) $396 + 12 = 408$

(C) $4{,}752 \times 12 = 57{,}024$

(D) $396 \div 12 = 33$

8. Which of the following is **not** a common factor of 24 and 36?

(A) 4

(B) 8

(C) 6

(D) 12

9. Riley's bank account balance is $-\$35$ and Jordan's bank account balance is $-\$20$. Who owes more money?

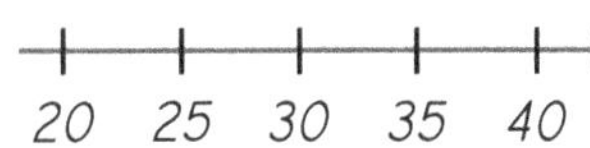

A Jordan, because $-20 > -35$

B Riley, because $|-35| = 35 > |-20| = 20$

C They owe the same amount

D Jordan, because $20 > 35$

10. **Part A:** In which quadrant is the point $(-8, 1)$ located?

Part B: If $(-8, 1)$ is reflected across the y-axis, what are the coordinates of its image and in which quadrant does it lie?

Part C: If the original point $(-8, 1)$ is instead reflected across the x-axis, what are the coordinates of its image and in which quadrant does it lie?

11. The number line below models a multiplication problem using repeated jumps.

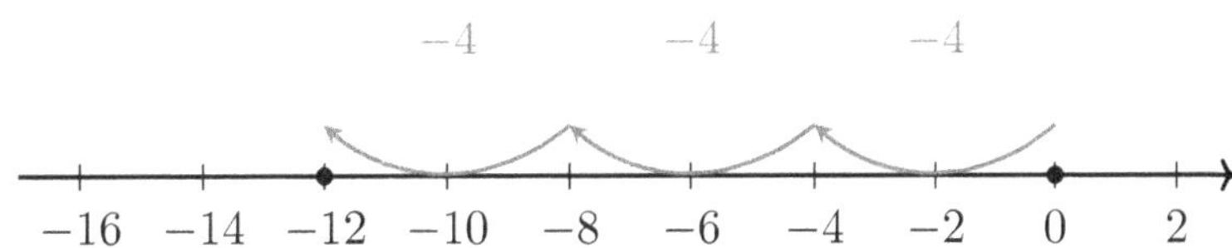

Which multiplication problem does this number line represent?

A $3 \times 4 = 12$

B $(-3) \times (-4) = 12$

C $3 \times (-4) = -12$

D $(-4) \times (-3) = 12$

12. Evaluate: $(5 - 1)^2 + 7$

A 14

B 23

C 11

D 33

13. You earn \$12 per hour and worked h hours. You also got a \$20 bonus. Write an expression for your total pay.

14. A student made a factor tree for the expression $4(x+3)$, shown below. Fill in the missing pieces labeled A and B.

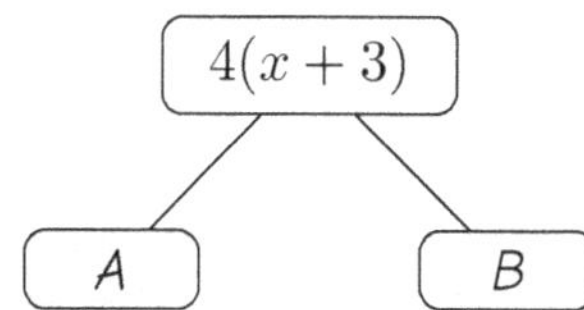

15. Evaluate $m^2 + 2m + 1$ when $m = 4$.

Your Answer:

16. Solve: $15a = 105$

17. Is $n = 4$ a solution to $n > 4$? Is $n = 4$ a solution to $n \geq 4$? Explain both.

18. Which of these numbers is a solution to $x \geq -3$?

   (A) -4 (B) -3.5

   (C) -3 (D) -100

Find more at
ViewMath.com/FL-Grade6

ViewMath.com

19. A bathtub drains at 3 gallons per minute. It starts with 45 gallons. Write an equation for the water w remaining after t minutes. When is the tub empty?

20. A student says the area of this parallelogram is 48 square units. The base is 8 and the slant side is 6. The actual height, shown by the dashed line, is 5. Is the student correct?

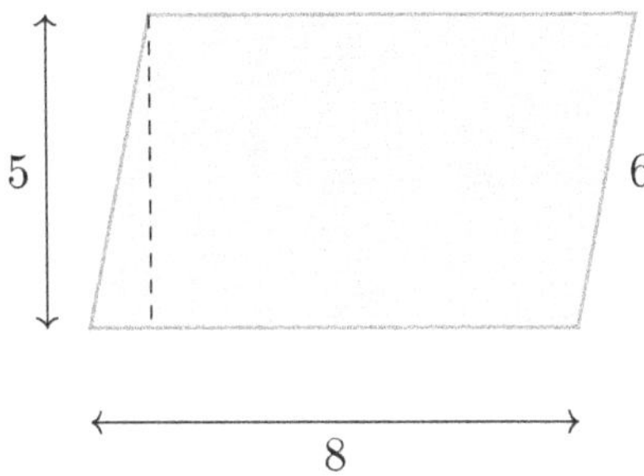

(A) Yes, because $8 \times 6 = 48$.

(B) No, the area is 40 because the height is 5, not 6.

(C) No, the area is 24 because you must divide by 2.

(D) Yes, because $\frac{1}{2} \times 8 \times 6 = 24$ and you double it.

21. A rectangular prism has length $\frac{1}{3}$ ft, width $\frac{1}{2}$ ft, and height 6 ft. What is the volume?

22. What is the distance between $(-4, 3)$ and $(6, 3)$?

(A) 2 units

(B) 10 units

(C) 6 units

(D) 7 units

Find more at
ViewMath.com/FL-Grade6

23. *What is a net?*

(A) *A 3D shape made of cubes*

(B) *A flat pattern that folds into a 3D shape*

(C) *The volume of a rectangular prism*

(D) *A grid used to measure area*

24. *Write a statistical question you could ask your classmates about food.*

Your Answer

25. *Data:* $50, 55, 60, 65, 70$. *The mean is 60. Find the MAD.*

Your Answer

26. *You survey 15 classmates about their favorite fruit. Which display should you start with?*

(A) *Histogram, because it groups data into intervals.*

(B) *Frequency table, because you need to count each category first.*

(C) *Box plot, because you need the five-number summary.*

(D) *Dot plot, because it shows individual numerical values.*

27. *A box plot has: min $= 60$, $Q1 = 70$, median $= 75$, $Q3 = 85$, max $= 100$. What percent of data falls between 70 and 85?*

Your Answer

28. *Which pair of measures should you report when outliers are present?*

(A) *Mean and range*

(B) *Mode and maximum*

(C) *Median and IQR*

(D) *Mean and IQR*

29. A bag contains 4 green marbles, 6 red marbles, and 10 blue marbles. What is the probability of randomly drawing a green marble? Write your answer as a percent.

30. Using the stem-and-leaf plot below, how many data values are greater than 50?

Stem	Leaves
3	2 5 8
4	0 4 7 9
5	1 3 6
6	2

Key: 3 | 2 means 32

(A) 3

(B) 4

(C) 5

(D) 6

 *End of Practice Test 8* 

Great job finishing the test!

 My Score

I got __________ out of 30 questions right.

Check your answers in the Answer Key at the back of the book

Review any questions you missed. That's how we learn!

Check Your Score Online!

Visit **ViewMath Academy** *to enter your answers and see which topics you need to review. You can also explore lessons, take quizzes, track your scores, and save your progress!*

viewmath.com/score/6.1.FL.23

Or go to viewmath.com/score and enter code: 6.1.FL.23

Practice Test 9

30 Questions

✏️ Before You Start ✏️

- ✔ **Read each question carefully** before choosing your answer.
- ✔ **Show your work** on scratch paper when you need to.
- ✔ **Skip hard questions** and come back to them later.
- ✔ **Check your answers** when you're done.
- ✔ **Take your time** — there's no rush!

⭐ *You've Got This!* ⭐

Do your best and show what you know!

1. The ratio of red to blue to green beads is $1 : 3 : 2$. There are 18 beads total. How many blue beads are there?

   (A) 3

   (B) 6

   (C) 9

   (D) 12

2. The ratio of red to blue beads is $1 : 4$. If there are 3 red beads, how many blue beads are there?

   (A) 4

   (B) 7

   (C) 12

   (D) 8

3. Two ratio graphs both pass through the origin. Line A goes through $(1, 4)$ and Line B goes through $(1, 3)$. Which line is steeper?

   (A) Line A

   (B) Line B

   (C) They have the same steepness.

   (D) Cannot be determined.

4. A test has 100 questions. Emma got 88 correct. What percent did she get right?

   (A) 12%

   (B) 8.8%

   (C) 88%

   (D) 78%

5. Tyler earns $80 from a weekend job. He spends $50 and saves the rest. What percent of his earnings does he save?

   (A) 25%

   (B) 30%

   (C) 37.5%

   (D) 62.5%

Find more at
ViewMath.com/FL-Grade6

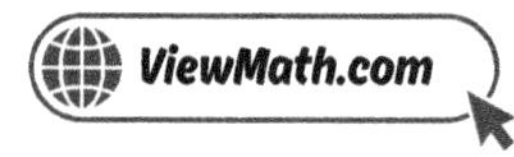

6. A ribbon is $\frac{3}{4}$ yard long. You cut pieces that are each $\frac{1}{8}$ yard. How many pieces can you cut?

(A) 6

(B) $\frac{3}{32}$

(C) 4

(D) 8

7. What is $2{,}688 \div 16$?

(A) 158

(B) 168

(C) 178

(D) 1,068

8. Marcus listed the factors of 48 and 72 in the table below, but left some blanks.

Factors of 48	1, 2, 3, 4, 6, 8, 12, 16, 24, 48
Factors of 72	1, 2, 3, 4, 6, 8, 9, 12, 18, 24, 36, 72

Part A: List all the common factors of 48 and 72.

Part B: What is the GCF of 48 and 72?

Part C: A baker made 48 cookies and 72 brownies. She wants to pack them into identical boxes with no treats left over. Using your answer from Part B, how many boxes can she fill, and what goes in each box?

9. Which statement about absolute value is **true**?

(A) Absolute value can be negative.

(B) $|-5| = -5$

(C) Absolute value is always greater than or equal to zero.

(D) $|0|$ is undefined.

10. *A triangle is drawn on the coordinate plane below with vertices at A, B, and C.*

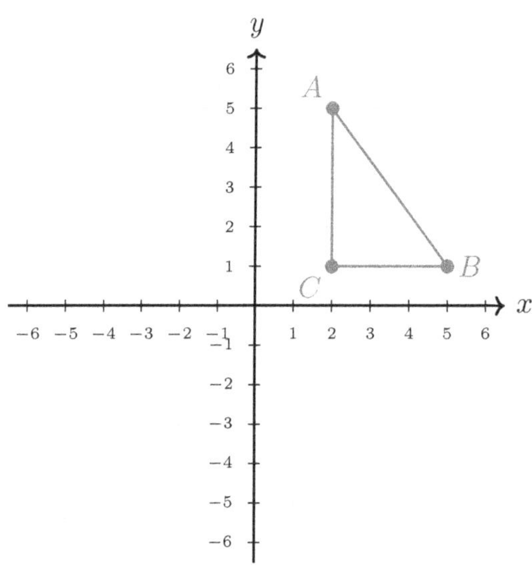

Part A: *Write the coordinates of each vertex: A, B, and C.*

Part B: *If the triangle is reflected across the y-axis, write the coordinates of the new vertices A′, B′, and C′.*

Part C: *In which quadrant will the reflected triangle be located?*

11. *What is $(-2) \times (-3) \times (-4)$?*

  (A) 24

  (B) -9

  (C) 9

  (D) -24

12. *What is the value of 4^3?*

  (A) 12

  (B) 43

  (C) 64

  (D) 81

13. Which expression represents "triple a number b, then subtract 4"?

(A) $3(b-4)$

(B) $3b-4$

(C) $b-12$

(D) $4-3b$

14. What is the coefficient of m in the expression $m+12$?

(A) 0

(B) 1

(C) 12

(D) m

15. Evaluate $a^2 + b^2$ when $a = 3$ and $b = 4$.

(A) 7

(B) 12

(C) 25

(D) 49

16. Solve: $\dfrac{t}{6} = 5$

(A) $t = 11$

(B) $t = 1$

(C) $t = 30$

(D) $t = 56$

17. Is $x = 5$ a solution to $x > 5$?

(A) Yes, because 5 is equal to 5

(B) Yes, because 5 is positive

(C) No, because 5 is not greater than 5

(D) No, because 5 is greater than 5

18. Which inequality matches a graph with an open circle at -2 and shading to the left?

(A) $x > -2$

(B) $x < -2$

(C) $x \geq -2$

(D) $x \leq -2$

Find more at
ViewMath.com/FL-Grade6

ViewMath.com

19. In $c = 7p$, where c is total cost and p is the number of pizzas, what is the cost of 6 pizzas?

   (A) $13

   (C) $42

   (B) $36

   (D) $67

20. A field is shaped like a parallelogram with base 25 m and height 14 m. Each bag of seed covers 50 m^2. How many bags are needed?

   (A) 5 bags

   (C) 7 bags

   (B) 6 bags

   (D) 8 bags

21. Two boxes have the same volume. Box A is $6 \times 4 \times 5$. Box B is $10 \times 3 \times h$. What is the height h of Box B?

   (A) 2

   (C) 6

   (B) 4

   (D) 8

22. What is the distance between the points $(2, 5)$ and $(2, -3)$?

   (A) 2 units

   (C) 8 units

   (B) 5 units

   (D) 3 units

23. A cube has a surface area of 96 cm^2. What is the edge length?

   (A) 3 cm

   (C) 6 cm

   (B) 4 cm

   (D) 8 cm

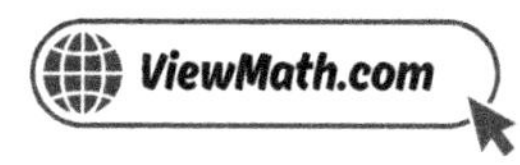

24. *Which question is **not** statistical?*

(A) *How many siblings do students in our school have?*

(B) *How far do students travel to get to school?*

(C) *What is 12×9?*

(D) *How many pets do families in our neighborhood own?*

25. *Look at the two dot plots below.*

Data Set A

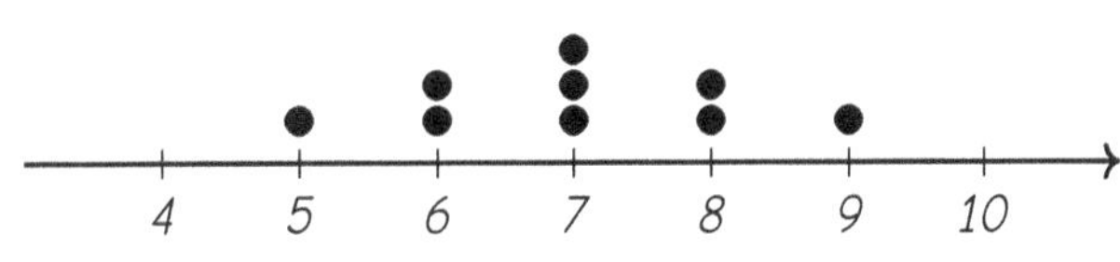

Data Set B

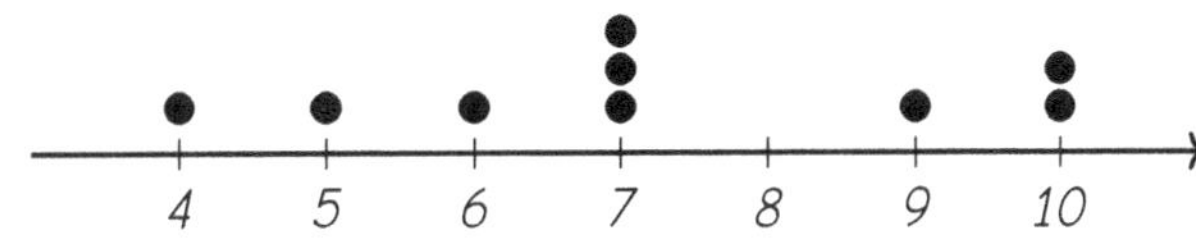

Which data set has a larger range?

(A) Data Set A

(B) Data Set B

(C) They have the same range.

(D) Cannot be determined.

26. *You want to know whether to display data as a dot plot, histogram, or frequency table. Which question should you ask first?*

(A) *What is the mean of the data?*

(B) *How many data points are there, and is the data numerical or categorical?*

(C) *What color should the bars be?*

(D) *Is the data symmetric or skewed?*

27. Look at the two box plots below showing test scores for two classes.

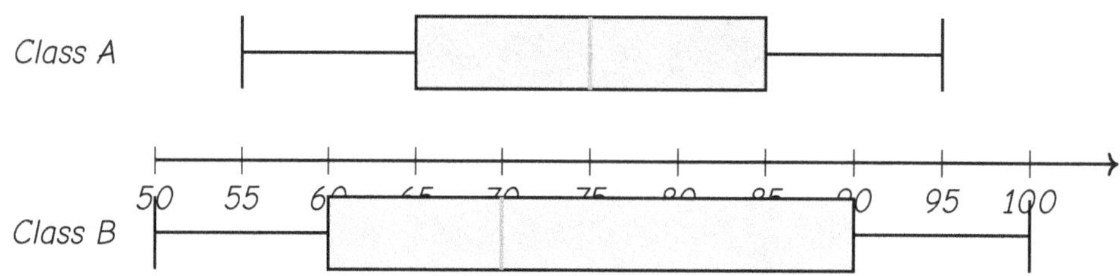

Which class has more consistent scores?

(A) Class A, because its box is narrower.

(B) Class B, because its box is wider.

(C) Both are equally consistent.

(D) Class B, because its median is higher.

28. Which measure of spread is affected by every single data value?

(A) IQR

(B) Range

(C) Median

(D) Mode

29. A jar holds 10 jellybeans: 4 cherry, 4 grape, and 2 lemon. Which event is **equally likely** as drawing a cherry jellybean?

(A) Drawing a lemon jellybean

(B) Drawing a grape jellybean

(C) Not drawing a jellybean

(D) Drawing a cherry or lemon jellybean

30. Find the mean of the data in the stem-and-leaf plot.

Stem	Leaves
2	0 4 8
3	2 6

Key: 2 | 0 means 20

Your Answer

Practice Test 10

✎ 30 Questions

✏ Before You Start ✏

- ✓ **Read each question carefully** before choosing your answer.
- ✓ **Show your work** on scratch paper when you need to.
- ✓ **Skip hard questions** and come back to them later.
- ✓ **Check your answers** when you're done.
- ✓ **Take your time** — there's no rush!

★ You've Got This! ★

Do your best and show what you know!

1. *The tape diagram below shows the ratio of apple juice to orange juice in a drink.*

Apple: ▢▢▢

Orange: ▢▢▢▢▢

If each part equals 4 ounces, how many total ounces of drink are there?

- A) 12
- B) 20
- C) 32
- D) 8

2. *The ratio table below shows the relationship between bags of seed and the area of garden they cover.*

Bags of Seed	Area (sq ft)
2	50
4	100
6	?
8	200

What is the missing value?

- A) 125
- B) 150
- C) 175
- D) 160

3. *Juice costs $2 per bottle. Which point would **not** appear on the graph of this ratio?*

- A) $(1, 2)$
- B) $(3, 6)$
- C) $(4, 10)$
- D) $(5, 10)$

Find more at
ViewMath.com/FL-Grade6

4. Order from least to greatest: $\frac{1}{3}$, 30%, 0.35.

(A) 30%, $\frac{1}{3}$, 0.35

(B) $\frac{1}{3}$, 30%, 0.35

(C) 0.35, 30%, $\frac{1}{3}$

(D) 30%, 0.35, $\frac{1}{3}$

5. Marcus has \$150 on a credit card at 10% annual interest and \$150 in a savings account earning 2% annual interest. After one year with no payments or withdrawals, what is the difference between the interest he **owes** and the interest he **earns**?

(A) \$8

(B) \$12

(C) \$15

(D) \$18

6. Bar A and Bar B below represent the same whole. Bar A has $\frac{2}{3}$ shaded. Bar B has $\frac{1}{6}$ shaded.

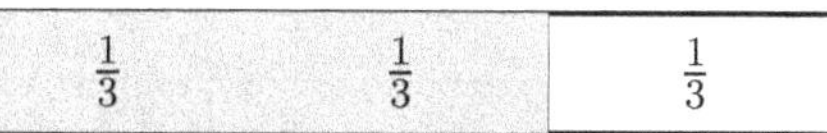

How many times does the shaded part of Bar B fit into the shaded part of Bar A? Write a division equation and solve.

Your Answer:

7. Compute $3{,}780 \div 15$.

Your Answer:

8. Two lights blink at regular intervals. One blinks every 4 seconds and the other blinks every 7 seconds. They both blink at time 0. After how many seconds will they blink at the same time again?

9. A number plus its opposite always equals which value?

   (A) 1   (B) The number itself

   (C) 0   (D) -1

10. In which quadrant is the point $(2, -5)$ located?

   (A) Quadrant I   (B) Quadrant II

   (C) Quadrant III   (D) Quadrant IV

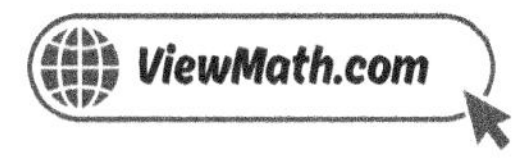

11. *Study the pattern table below.*

Expression	Product
$(-3) \times 3$	-9
$(-3) \times 2$	-6
$(-3) \times 1$	-3
$(-3) \times 0$	0
$(-3) \times (-1)$	?
$(-3) \times (-2)$	?
$(-3) \times (-3)$	?

Part A: *Complete the three missing products by continuing the pattern.*

Part B: *Explain the pattern and how it shows that a negative times a negative is positive.*

12. *What is the value of 1^{10}?*

(A) 0

(B) 1

(C) 10

(D) 100

13. *Look at the balance model below. Each triangle represents the same unknown number x, and each small square represents 1. Write an expression for the total value shown on the balance.*

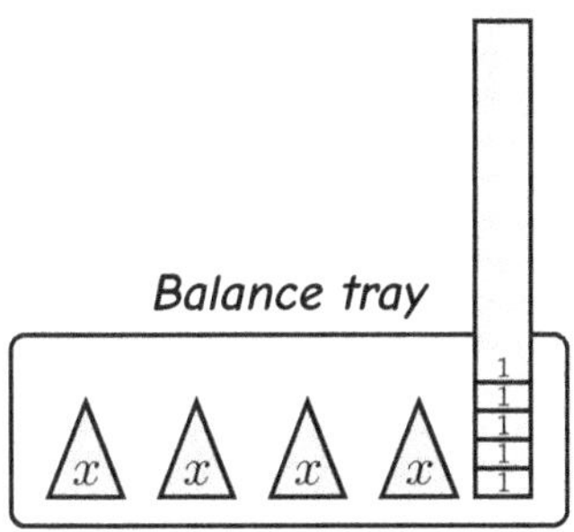

14. *Which statement is true about the expression $x + 5$?*

(A) *The coefficient of x is 0*   (B) *There are 3 terms*

(C) *5 is the coefficient of x*   (D) *The coefficient of x is 1*

15. *Evaluate $10 - 2(k - 1)$ when $k = 4$.*

(A) 2   (B) 4

(C) 6   (D) 14

16. *Solve: $m + 17 = 30$*

17. *Which inequality represents "you need more than \$25 to buy the video game"?*

(A) $d \leq 25$ (B) $d < 25$

(C) $d > 25$ (D) $d \geq 25$

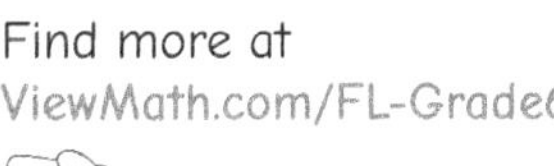
Find more at
ViewMath.com/FL-Grade6

ViewMath.com

18. Look at the number line below and write the inequality it represents.

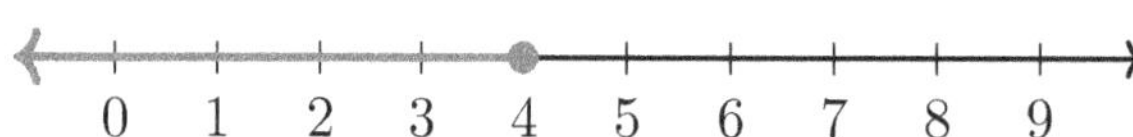

Your Answer

19. Which equation represents this relationship: "The number of legs L is 4 times the number of dogs d"?

(A) $d = 4L$

(B) $L = d + 4$

(C) $L = 4d$

(D) $L = d \div 4$

20. A trapezoid has bases $b_1 = 5$ ft and $b_2 = 13$ ft, and height $h = 8$ ft. What is the area?

(A) $144 \ ft^2$

(B) $72 \ ft^2$

(C) $52 \ ft^2$

(D) $36 \ ft^2$

21. What is the volume of a rectangular prism with length 5 cm, width 3 cm, and height 4 cm?

(A) $12 \ cm^3$

(B) $60 \ cm^3$

(C) $30 \ cm^3$

(D) $94 \ cm^3$

22. A triangle on the coordinate plane has vertices $(0,0)$, $(8,0)$, and $(4,6)$. What is the length of the base along the x-axis?

(A) 4 units

(B) 6 units

(C) 8 units

(D) 10 units

Find more at
ViewMath.com/FL-Grade6

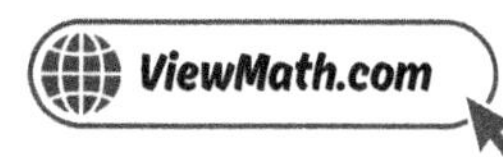

23. A triangular prism has two equilateral triangle bases with side 6 cm and height 5.2 cm. The prism is 12 cm long. What is the total surface area?

24. Look at the flowchart below. Which path correctly identifies a statistical question?

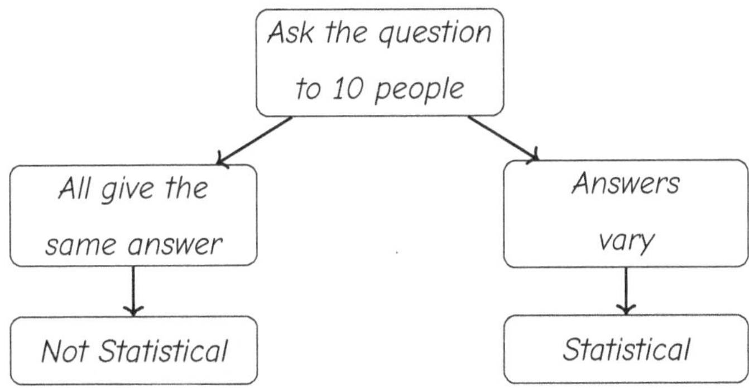

Using the flowchart, which question would end at "Statistical"?

(A) How many wheels does a bicycle have?   (B) What is the sum of $10 + 5$?

(C) How many minutes do you spend eating   (D) How many days are in one week?
lunch?

25. If every value in a data set is increased by 5, what happens to the range?

(A) It increases by 5.   (B) It decreases by 5.

(C) It stays the same.   (D) It doubles.

26. A histogram shows minutes of exercise: 0–14 (3), 15–29 (8), 30–44 (12), 45–59 (7). How many people exercised less than 30 minutes?

Find more at
ViewMath.com/FL-Grade6

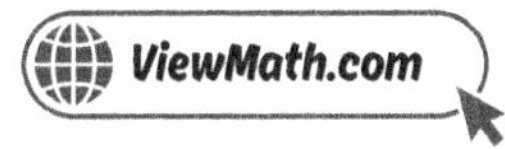

27. In a box plot, the "box" stretches from —

(A) the minimum to the maximum   (B) Q1 to Q3

(C) the median to the maximum   (D) 0 to the median

28. Two box plots are shown on the same number line.

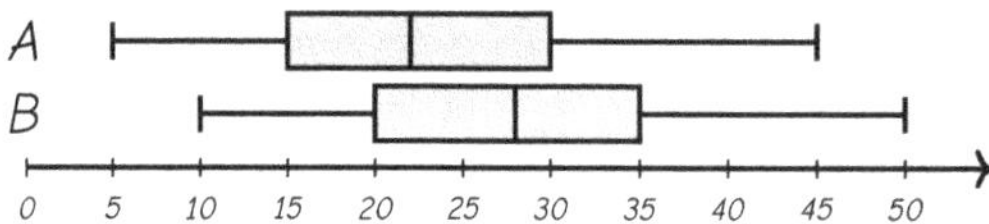

Which data set has a greater IQR?

(A) A (IQR = 15)   (B) B (IQR = 15)

(C) They have the same IQR.   (D) Cannot be determined.

29. A bag has 5 orange balls, 3 purple balls, and 2 yellow balls. What is the probability of **not** drawing a yellow ball?

(A) $\dfrac{2}{10}$ (B) $\dfrac{5}{10}$

(C) $\dfrac{8}{10}$ (D) $\dfrac{3}{10}$

30. Which stem has the most data values?

Stem	Leaves
4	0 2
5	1 3 5 7 9
6	2 6 8
7	4

Key: 4 | 0 means 40

(A) 4   (B) 5

(C) 6   (D) 7

 End of Practice Test 10 

Great job finishing the test!

My Score

I got _______________ out of 30 questions right.

Check your answers in the Answer Key at the back of the book.

Review any questions you missed. That's how we learn!

Check Your Score Online!

Visit **ViewMath Academy** to enter your answers and see which topics you need to review. You can also explore lessons, take quizzes, track your scores, and save your progress!

viewmath.com/score/6.1.FL.25

Or go to viewmath.com/score and enter code 6.1.FL.25

Answer Key & Explanations

Answer Key

First try each test on your own, then check your work here.

✅ Practice Test 1 — Answer Key

1 5   **2** C   **3** B   **4** D   **5** C   **6** 4   **7** C   **8** A

9 $|-120| = 120,\ |-85| = 85$; *Store A is farther from break-even.*   **10** $(-7, -3)$ **11** C

12 $3^3 = 27$ *unit cubes*   **13** *5 times a number* x, *plus 3 (or equivalent wording)*   **14** B   **15** $23

16 C   **17** *Answers vary. Example: A classroom has at most 15 computers.*

18 $x = 3$ *IS a solution to* $x \le 3$ *because* $3 \le 3$ *is true.* $x = 3$ *is NOT a solution to* $x < 3$ *because* $3 < 3$ *is false.*

19 B   **20** A   **21** C   **22** D   **23** B   **24** B   **25** D   **26** B   **27** B

28 *Not always.*   **29** B   **30** 45

💡 Time to Learn! 💡

*Review the explanations below, **especially for the questions you missed**.*

Understanding why each answer is correct builds stronger problem-solving skills.

Tip: *Circle any questions you got wrong, then read their explanation carefully.*

Find more at
ViewMath.com/FL-Grade6

📖 Practice Test 1 — Detailed Explanations

1. Each part $= 20 \div 4 = 5$. Vegetables $= 1 \times 5 = 5$.

2. From the graph, each hour earns $10 (2 units on the y-axis $=$ $10). In 6 hours: $6 \times 10 = 60.

3. Ratio: $1 : 4$. When $x = 3$: $y = 3 \times 4 = 12$.

4. Divide by 100: $72 \div 100 = 0.72$.

5. The table shows that only the credit card does not use your own money and can charge interest, meaning it involves borrowing.

6. $2\frac{1}{2} = \frac{5}{2}$. Then $\frac{5}{2} \div \frac{5}{8} = \frac{5}{2} \times \frac{8}{5} = \frac{40}{10} = 4$ shelves.

7. Total cans: $384 + 528 + 456 + 432 = 1{,}800$. Divide: $1{,}800 \div 12 = 150$ cans per classroom.

8. Multiples of 4: $4, 8, 12, 16, 20, \ldots$ Multiples of 10: $10, 20, \ldots$ The smallest number in both lists is 20.

9. $|-120| = 120$ and $|-85| = 85$. Since $120 > 85$, Store A was farther from break-even. Store A's larger absolute value means its loss was bigger.

10. Reflecting across the x-axis changes the sign of the y-coordinate: $(x, y) \to (x, -y)$. So $(-7, 3)$ becomes $(-7, -3)$.

11. The machine multiplies by -6. Input -7: $(-7) \times (-6) = 42$ (same signs $\to$ positive).

12. Each edge is 3 units. The cube needs $3 \times 3 \times 3 = 3^3 = 27$ unit cubes.

Find more at
ViewMath.com/FL-Grade6

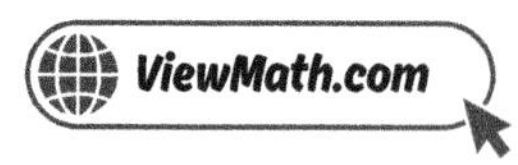

13 $5x$ means "5 times x" and $+3$ means "plus 3."

14 The terms are $4x$, 9, and $2y$. Terms are separated by $+$ or $-$ signs.

15 $5 + 3(6) = 5 + 18 = 23$ dollars.

16 Multiply both sides by 4: $m = 7 \times 4 = 28$.

17 Any situation where a quantity is 15 or fewer works.

18 $\leq$ includes the boundary value; $<$ does not.

19 In B, y is always 7 no matter what x is. The value of y does not change with x.

20 Rectangle: $8 \times 5 = 40$. Triangle: $\frac{1}{2} \times 8 \times 3 = 12$. Total: $40 + 12 = 52$ cm^2.

21 Volume measures 3-dimensional space, so it uses cubic units like cm^3, in^3, or ft^3.

22 $|4 - (-5)| = |9| = 9$ units.

23 $SA = 2(10)(3) + 2(10)(7) + 2(3)(7) = 60 + 140 + 42 = 242$ m^2.

24 The answers range from 0 to 4 and vary from person to person, which confirms it is a statistical question.

25 Range $= 35 - 5 = 30$.

26 1 hour has 6 dots (highest frequency) ⊠ mode. 8 hours is far from the cluster (0–4) and has only 1 dot ⊠ outlier.

Find more at
ViewMath.com/FL-Grade6

27. The left whisker extends from the minimum to $Q1$, covering the lowest 25% of the data.

28. A single outlier can make the range very large while the rest of the data is tightly clustered. The IQR gives a better picture of overall spread. For example, $\{1, 50, 51, 52, 53\}$ has range 52 but most values are close together.

29. Numbers ≤ 3: $1, 2, 3$. That is 3 out of 6. $P(\leq 3) = \dfrac{3}{6} = \dfrac{1}{2}$.

30. The value 45 appears 3 times (three leaves of 5 on stem 4). No other value appears more often. The mode is 45.

☑ Practice Test 2 — Answer Key

1. C   2. 12, 16, and 30   3. Part A: Runner A $= 3 : 4$; Runner B $= 2 : 3$.   Part B: Runner A is faster.

4. C   5. C   6. D   7. A   8. GCF $= 4$;   LCM $= 24$ 9. B

10. The point $(9, 0)$ lies on the x-axis because its y-coordinate is 0. Points on the axes are not inside any quadrant.

11. C   12. 4   13. $3n - 8$ 14. 12 and k 15. C   16. B   17. C   18. C

19. Independent: h (hours). Dependent: d (distance).   20. B   21. B   22. 26 units   23. C

24. No   25. C   26. Total $= 100$, Mode $=$ Bus, Walk $= 15\%$ 27. B   28. A   29. C   30. B

💡 Time to Learn! 💡

Review the explanations below, **especially for the questions you missed**.

Understanding why each answer is correct builds stronger problem-solving skills.

Tip: Circle any questions you got wrong, then read their explanation carefully.

Find more at
ViewMath.com/FL-Grade6

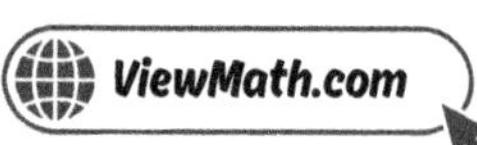

Practice Test 2 — Detailed Explanations

Girls have 6 parts $= 24$, so each part $= 24 \div 6 = 4$. Boys have 4 parts $= 4 \times 4 = 16$.

Row 2: $4 \times 2 = 8$, so $6 \times 2 = 12$. Row 3: $6 \times 4 = 24$, so $4 \times 4 = 16$. Row 4: $4 \times 5 = 20$, so $6 \times 5 = 30$.

Runner A: 3 laps in 4 min $= 0.75$ laps/min. Runner B: 2 laps in 3 min ≈ 0.667 laps/min. $0.75 > 0.667$, so Runner A is faster. On the graph, Runner A's line is steeper.

$50\% = \dfrac{50}{100} = \dfrac{1}{2}$.

Savings: $0.20 \times \$240 = \$48$. Supplies: $0.10 \times \$240 = \$24$. Left: $\$240 - \$48 - \$24 = \$168$.

$\dfrac{7}{8} \div \dfrac{1}{4} = \dfrac{7}{8} \times \dfrac{4}{1} = \dfrac{28}{8} = \dfrac{7}{2} = 3\dfrac{1}{2}$ laps.

$7 \div 3 = 2$ R1; bring down $5 \to 15 \div 3 = 5$; bring down $6 \to 6 \div 3 = 2$. Answer: 252. Check: $252 \times 3 = 756$.

GCF: Factors of 8: $1, 2, 4, 8$. Factors of 12: $1, 2, 3, 4, 6, 12$. Common factors: $1, 2, 4$. GCF $= 4$. LCM: Multiples of 8: $8, 16, 24, \ldots$ Multiples of 12: $12, 24, \ldots$ LCM $= 24$

Find each absolute value: $|3| = 3$, $|-8| = 8$, $|-1| = 1$, $|5| = 5$. The greatest absolute value is 8, which belongs to -8.

The four quadrants are the regions between the axes. A point is on an axis when $x = 0$ (on the y-axis) or $y = 0$ (on the x-axis). Since $(9, 0)$ has $y = 0$, it sits on the x-axis itself, which is the boundary between Quadrant I and Quadrant IV—not in either one.

Losing 4 points is -4 per round. Over 9 rounds: $9 \times (-4) = -36$ points.

Find more at
ViewMath.com/FL-Grade6

12. *Parentheses: $12 - 4 = 8$. Exponent: $8^2 = 64$. Divide: $64 \div 16 = 4$.*

13. *Start with n. Multiply by 3: $3n$. Subtract 8: $3n - 8$.*

14. *$12k = 12 \times k$. The factors are 12 and k.*

15. *$(10 + 6) \div 2 = 16 \div 2 = 8$.*

16. *Subtract 15: $k = 32 - 15 = 17$.*

17. *"Fewer than 20" means less than 20: $y < 20$.*

18. *Open circle at 0 means 0 is not included. Shading right means greater than: $x > 0$.*

19. *You choose the hours (independent). Distance depends on hours: $d = 6h$.*

20. *Area uses the height, not the slant side. $A = 12 \times 6 = 72 \ m^2$.*

21. *Total volume $= 5 \times 3 \times 2 = 30 \ m^3$. Half-full: $30 \div 2 = 15 \ m^3$.*

22. *Length $= |2 - (-6)| = 8$. Width $= |2 - (-3)| = 5$. Perimeter $= 2(8) + 2(5) = 26$ units.*

23. *A rectangular prism has 6 faces: 3 pairs of opposite rectangles.*

24. *"How many inches in a foot?" has a numerical answer (12) but is not statistical because the answer does not vary. A statistical question must produce data that varies.*

25. *$IQR = Q3 - Q1 = 55 - 35 = 20$.*

Find more at
ViewMath.com/FL-Grade6

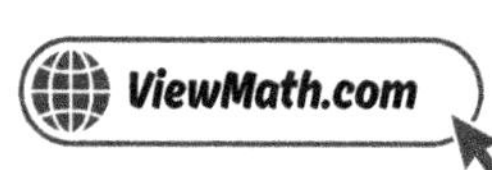

26. $45 + 30 + 15 + 10 = 100$. Bus has the highest frequency ⇒ mode. Walk: $15/100 = 15\%$.

27. The five-number summary consists of minimum, first quartile (Q1), median, third quartile (Q3), and maximum.

28. Same medians, but Athlete A's range (1.2) is smaller — the times vary less, indicating more consistency.

29. Event Y is positioned closest to 0.75 on the probability scale.

30. Class A scores: $62, 65, 68, 71, 74, 77, 80, 86$ (8 values). Median $= (71 + 74) \div 2 = 72.5$. Class B scores: $63, 66, 70, 72, 75, 79, 81, 84, 92$ (9 values). Median $= 75$ (the 5th value). Class B has the higher median.

✅ Practice Test 3 — Answer Key

1 C	2 C	3 B	4 C	5 C	6 B	7 C	8 A	9 B	10 B
11 B	12 8	13 C	14 1	15 A	16 A	17 D	18 B	19 B	20 D
21 A	22 30 units	23 A	24 B	25 B	26 Mode = 5000, Range = 4000			27 B	
28 B	29 C	30 B							

💡 Time to Learn! 💡

Review the explanations below, **especially for the questions you missed**.

Understanding why each answer is correct builds stronger problem-solving skills.

Tip: Circle any questions you got wrong, then read their explanation carefully.

Find more at
ViewMath.com/FL-Grade6

📖 Practice Test 3 — Detailed Explanations

1. The question asks flour to eggs. Flour $= 3$, eggs $= 2$. The ratio is $3 : 2$.

2. $6 \times 3 = 18$ cups of flour, so $4 \times 3 = 12$ eggs.

3. The ratio is $6 : 2 = 3 : 1$. When $x = 15$: $y = 15 \div 3 = 5$.

4. 50% of $200 = \dfrac{1}{2} \times 200 = 100$ items.

5. Interest $= \$500 \times 0.18 = \$90$. New balance $= \$500 + \$90 = \$590$.

6. $\dfrac{3}{5} \times \dfrac{5}{1} = \dfrac{15}{5} = 3$.

7. $73 \div 24 = 3$ R1; bring down $4 \to 14 \div 24 = 0$ R14; bring down $4 \to 144 \div 24 = 6$. Answer: 306. The zero in the tens place must not be skipped. Check: $306 \times 24 = 7{,}344$.

8. Factors of 36: $1, 2, 3, 4, 6, 9, 12, 18, 36$. Factors of 48: $1, 2, 3, 4, 6, 8, 12, 16, 24, 48$. Common factors: $1, 2, 3, 4, 6, 12$. The greatest is 12.

9. The manager found the distance from zero, which is the absolute value. $|-200| = 200$, so the company is $\$200$ from break-even. Absolute value answers "how far from zero?"

10. In Quadrant II, the x-coordinate is negative (left of the y-axis) and the y-coordinate is positive (above the x-axis), giving the sign convention $(-, +)$.

11. Count the negative factors: there are 4 (even number), so the product is positive. $1 \times 2 \times 5 \times 3 = 30$.

Parentheses: $3 + 2 = 5$. Exponent: $2^2 = 4$. Divide: $60 \div 5 = 12$. Subtract: $12 - 4 = 8$.

"Subtract w from 15" means start with 15 and take away w: $15 - w$.

w means $1 \cdot w$. The coefficient is 1 even though it is not written.

$2(3(5) - 1) = 2(15 - 1) = 2(14) = 28$.

Subtract 24: $x = 50 - 24 = 26$.

"At most" means the value can equal the number or be less, so it should be $\leq$, not $<$.

"More than 75" is $t > 75$: open circle (not including 75), shade right (greater values).

2 cups per batch, b batches: $f = 2b$.

$A = \frac{1}{2}(40 + 60)(30) = \frac{1}{2}(100)(30) = 1,500 \ m^2$.

$V = \frac{3}{4} \times 2 \times 4 = \frac{3}{4} \times 8 = 6 \ ft^3$.

Length $= |4 - (-5)| = 9$. Width $= |4 - (-2)| = 6$. Perimeter $= 2(9) + 2(6) = 30$ units.

$SA = 2(5)(3) + 2(5)(4) + 2(3)(4) = 30 + 40 + 24 = 94 \ cm^2$. Yes, it is correct.

There are always 3 feet in one yard. This is not a statistical question because the answer does not vary.

MAD (Mean Absolute Deviation) is calculated by finding the average of the distances of each data value from the mean.

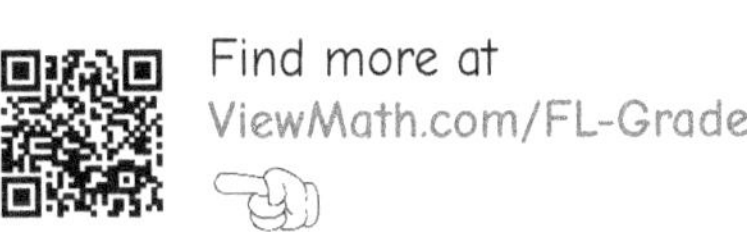
Find more at
ViewMath.com/FL-Grade6

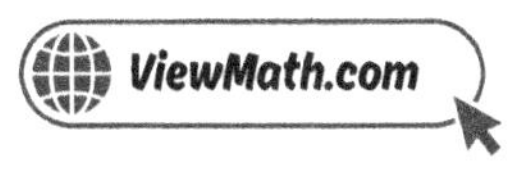

26 5000 has frequency 6 (highest). Range $= 7000 - 3000 = 4000$.

27 10 values. Median $= (9 + 11) \div 2 = 10$. Lower half: $1, 3, 5, 7, 9$. $Q1 = 5$. Upper half: $11, 13, 15, 17, 19$. $Q3 = 15$. $IQR = 15 - 5 = 10$.

28 A small IQR means the middle 50% is tightly packed. A large range means the minimum and maximum are far apart, indicating extreme values stretch the data.

29 There are 6 equally likely outcomes. Only 1 of them is a 4. $P(4) = \dfrac{1}{6}$.

30 The stem 11 combined with leaf 5 gives the three-digit number 115.

✅ Practice Test 4 — Answer Key

1 B   **2** 64   **3** C   **4** 87.5%; $\dfrac{7}{8}$ **5** C   **6** C   **7** B   **8** 16 bags   **9** B

10 B   **11** C   **12** $9^4 = 6,561$ **13** B   **14** 5 and $(y + 8)$ **15** C   **16** A   **17** D

18 B   **19** C   **20** C   **21** A   **22** C   **23** A

24 The teacher has one specific age, so the answer does not vary.   **25** C   **26** B   **27** B   **28** A

29 A: 0 (impossible); B: $\dfrac{1}{2}$ (equally likely); C: $\dfrac{3}{4}$ (likely); D: 1 (certain)   **30** B

Find more at
ViewMath.com/FL-Grade6

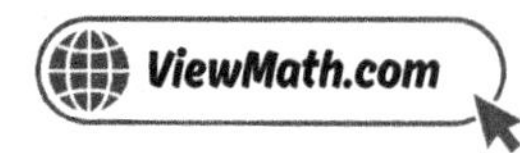

> ### 💡 Time to Learn! 💡
>
> *Review the explanations below, **especially for the questions you missed**.*
>
> *Understanding why each answer is correct builds stronger problem-solving skills.*
>
> ***Tip:*** *Circle any questions you got wrong, then read their explanation carefully.*

📖 Practice Test 4 — Detailed Explanations

"3 pencils for every 1 eraser" means pencils to erasers is $3 : 1$.

$3 \times 8 = 24$ red, so white $= 5 \times 8 = 40$. Total $= 24 + 40 = 64$.

$5 : 15$ simplifies to $1 : 3$.

$0.875 \times 100 = 87.5\%$. $0.875 = \dfrac{875}{1000} = \dfrac{7}{8}$.

Groceries are food, which is a basic need. The other choices are things you might enjoy but can live without.

$\dfrac{2}{3} \times \dfrac{5}{4} = \dfrac{10}{12} = \dfrac{5}{6}$.

The four steps of the standard algorithm are **Divide, Multiply, Subtract, Bring Down**. After dividing, you multiply the partial quotient by the divisor before subtracting.

Find the GCF of 48 and 80. Factors of 48: $1, 2, 3, 4, 6, 8, 12, 16, 24, 48$. Factors of 80: $1, 2, 4, 5, 8, 10, 16, 20, 40, 80$. GCF $= 16$. She makes 16 bags, each with 3 pencils and 5 erasers.

Distance from \$0 is measured by absolute value. $|-18| = 18$ and $|-25| = 25$. Since $18 < 25$, Mei's balance is closer to \$0. The smaller the absolute value, the closer to zero.

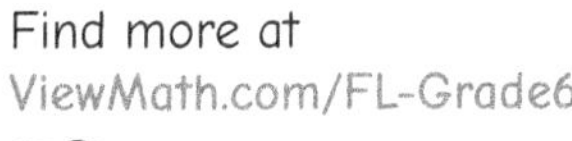

Find more at
ViewMath.com/FL-Grade6

ViewMath.com

10. Quadrant II has sign convention $(-,+)$. Since $-4 < 0$ and $7 > 0$, the point $(-4, 7)$ is in Quadrant II. A common mistake is choosing Quadrant III, which has both coordinates negative.

11. Each day the stock loses \$3, represented as -3. Over 5 days: $5 \times (-3) = -15$. The total change is $-\$15$.

12. $9^4 = 9 \times 9 \times 9 \times 9 = 6{,}561$.

13. "8 less than a number" means subtract 8 from n: $n - 8$.

14. $5(y + 8) = 5 \times (y + 8)$. The factors are 5 and $(y + 8)$.

15. $3^2 + 3 = 9 + 3 = 12$.

16. Start at unknown x, add 8, arrive at 14: $x + 8 = 14$, so $x = 6$.

17. $2.5 \geq 3$ is false. $2.5 < 3$, so it is not a solution.

18. $\leq$ means 5 is included (closed circle). Less than 5 is to the left (shade left).

19. $i = 12(5) = 60$ inches.

20. The height of a parallelogram is the perpendicular distance from the base to the opposite side, not the slanted side.

21. Base area $= 10 \times 4 = 40$. Height $= 120 \div 40 = 3$ cm.

22. Length $= |3 - (-5)| = 8$. Width $= |4 - (-2)| = 6$. Perimeter $= 2(8) + 2(6) = 28$ units.

23. Net A is a valid cross-shaped cube net. Net B has two squares on the same side, which causes overlap. Net C forms a 2×3 block, which doesn't fold into a cube.

24. A statistical question expects data that varies. The teacher's age is a single fixed number.

25. Upper half: $12, 14, 16$. The median of the upper half is 14, so $Q3 = 14$.

26. Range $= 74 - 68 = 6$.

27. Short whiskers and a narrow box indicate small range and small IQR, meaning data values are close together.

28. $Q_1 = (15 + 18)/2 = 16.5$. $Q_3 = (25 + 28)/2 = 26.5$. $IQR = 26.5 - 16.5 = 10$.

29. Event A: A standard die has faces 1–6, so rolling a 7 is impossible; $P = 0$. Event B: $P(\text{tails}) = \dfrac{1}{2}$. Event C: $P(\text{red}) = \dfrac{6}{8} = \dfrac{3}{4} = 0.75$. Event D: All faces (1–6) are less than 7, so $P = 1$ (certain).

30. Stem 3: leaves $2, 5$ (2 leaves). Stem 4: leaves $1, 4, 8$ (3 leaves). Stem 5: leaves $3, 7$ (2 leaves). Stem 4 has the most.

✓ Practice Test 5 — Answer Key

 10    B    B    D    C    $\dfrac{7}{2}$ or $3\dfrac{1}{2}$  C    A   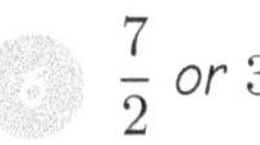 B

 $(3, -2)$  -63  B   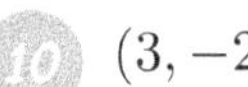 $\dfrac{t}{4} - 1$  C    A    $w = 72$  C

 C    C    B    $60\ \text{cm}^3$  B    A    B   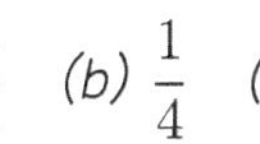 A    C

27. Range $= 55$, $IQR = 30$ 28. B   29. (a) $\dfrac{3}{8}$ (b) $\dfrac{1}{4}$ (c) $\dfrac{5}{8}$ 30. B

Find more at
ViewMath.com/FL-Grade6

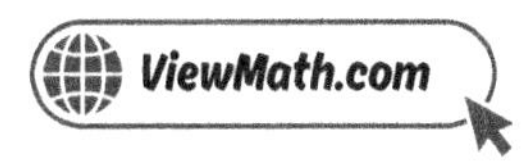

💡 Time to Learn! 💡

*Review the explanations below, **especially for the questions you missed**.*

Understanding why each answer is correct builds stronger problem-solving skills.

***Tip:** Circle any questions you got wrong, then read their explanation carefully.*

📖 Practice Test 5 — Detailed Explanations

1. *Total parts* $= 3 + 2 = 5$. *Each part* $= 25 \div 5 = 5$. *Swimming* $= 2 \times 5 = 10$.

2. $2 : 3$ *multiplied by 3 gives* $6 : 9$. *The other pairs do not simplify to* $2 : 3$.

3. *Ratio* $= 3 : 2$. *Double:* $(6, 4)$. *Triple:* $(9, 6)$, *not* $(9, 4)$ *or* $(9, 8)$.

4. *Zoo* $= 60\%$. *Remaining* $= 40\%$. *Farm is twice Park, so Farm* $= 2x$ *and Park* $= x$, *giving* $3x = 40\%$, $x \approx 13.3\%$. *Farm* $\approx 26.7\% \approx 27\%$.

5. *A credit card lets you borrow money from the card company. If you do not pay on time, you are charged interest.*

6. $\dfrac{7}{8} \div \dfrac{1}{4} = \dfrac{7}{8} \times \dfrac{4}{1} = \dfrac{28}{8} = \dfrac{7}{2} = 3\dfrac{1}{2}$ *pieces.*

7. $15 \div 5 = 3$; *bring down* $7 \to 7 \div 5 = 1$ *R2; bring down* $5 \to 25 \div 5 = 5$. *Answer:* 315. *Check:* $315 \times 5 = 1{,}575$.

8. *Find the GCF of 72 and 96. Factors of 72:* $1, 2, 3, 4, 6, 8, 9, 12, 18, 24, 36, 72$. *Factors of 96:* $1, 2, 3, 4, 6, 8, 12, 16, 24, 32, 48, 96$. *Common factors include* $1, 2, 3, 4, 6, 8, 12, 24$. *The greatest is 24 inches.*

$|-22| = 22$ and $|-15| = 15$. Since $22 > 15$, Beth's debt is larger. Note that $-15 > -22$ on the number line, but that does not mean Amy owes more — it means Amy's balance is less negative. Absolute value compares the size of each debt.

Start at $(0,0)$. Move 4 left and 3 up: $(-4, 3)$. Then move 7 right: $-4 + 7 = 3$, and 5 down: $3 - 5 = -2$. The final point is $(3, -2)$.

Different signs $\rightarrow$ negative. $9 \times 7 = 63$, so $(-9) \times 7 = -63$.

Multiply first: $5 \times 2 = 10$. Then add: $3 + 10 = 13$.

Divide t by 4 to get $\frac{t}{4}$, then subtract 1: $\frac{t}{4} - 1$.

$5m - 2n + 7 + m$ has 4 terms: $5m$, $2n$, 7, and m.

$20 - 3(4) = 20 - 12 = 8$.

Multiply by 8: $w = 9 \times 8 = 72$.

$t \leq 30$ means 30 or less, which is "no more than 30 degrees."

Closed circle means 3 is included ($\geq$ or $\leq$). Shading right means greater values: $x \geq 3$.

You choose x (independent), and y is computed from it (dependent). y increases as x increases.

When both bases are equal ($b_1 = b_2 = 4$), the trapezoid is actually a parallelogram. Area $= \frac{1}{2}(4 + 4)(6) = 24 = 4 \times 6$.

$V = 2.5 \times 4 \times 6 = 60 \ cm^3$.

Find more at
ViewMath.com/FL-Grade6

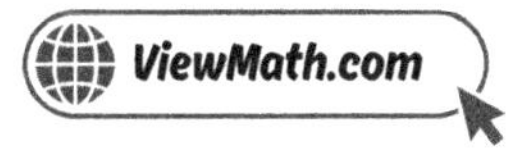

22 Horizontal side: $|7 - 1| = 6$. Vertical side: $|2 - (-4)| = 6$. Both sides are 6 units, so it is a square. The side length is 6.

23 $SA = 2(5)(5) + 2(5)(10) + 2(5)(10) = 50 + 100 + 100 = 250\ in^2$.

24 Different sixth graders spend different amounts of time on homework, so the answers vary. The other questions all have a single fixed answer.

25 The mean is 10. Every value equals 10, so every distance from the mean is 0. $MAD = 0 \div 5 = 0$.

26 Since two values share the highest frequency of 4, the data set is bimodal — it has 2 modes.

27 Range $= 80 - 25 = 55$. $IQR = 65 - 35 = 30$.

28 Same median (same typical sales). Store A's range (\$200) is much smaller than Store B's (\$800), making Store A more consistent.

29 There are 8 equal sections: 3 red, 2 blue, 3 green. (a) $P(red) = \dfrac{3}{8}$. (b) $P(blue) = \dfrac{2}{8} = \dfrac{1}{4}$. (c) $P(not\ green) = 1 - \dfrac{3}{8} = \dfrac{5}{8}$.

30 The leaves for stem 5 should be written 1 3 7 9, not 3 1 7 9. Leaves must always go from least to greatest.

✓ Practice Test 6 — Answer Key

1 Bananas to yogurt: $3 : 4$;   Yogurt to bananas: $4 : 3$ 2 B   3 A   4 C   5 B   6 B

7 B   8 A   9 D   10 C   11 B   12 B   13 $\dfrac{a+b}{2}$ 14 D   15 C   16 A

Find more at
ViewMath.com/FL-Grade6

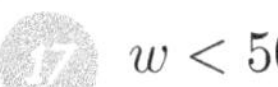 $w < 50$  Solutions: 0 and 5 (or any values greater than -1). Non-solution: -1 (or any value ≤ -1)

 B    $48\ in^2$  B    12 units    A    C    B    C    D

 B    C    B

💡 Time to Learn! 💡

*Review the explanations below, **especially for the questions you missed**.*

Understanding why each answer is correct builds stronger problem-solving skills.

Tip: *Circle any questions you got wrong, then read their explanation carefully.*

📖 Practice Test 6 — Detailed Explanations

 "3 bananas for every 4 cups of yogurt" gives $3 : 4$. Flip the order for yogurt to bananas: $4 : 3$.

 $3 \times 2 = 6$, so $5 \times 2 = 10$.

 The ratio is $2 : 3$. Choice A continues with $6 : 9 = 2 : 3$. Choice B has $6 : 8$ which is not $2 : 3$.

 Multiply by 100: $0.35 \times 100 = 35\%$.

Saving regularly builds a fund for emergencies and future goals like college, a car, or unexpected bills.

The reciprocal of a fraction is found by swapping the numerator and denominator:  $\frac{3}{7} \rightarrow \frac{7}{3}$.

$2{,}016 \div 14 = 144$. Check: $144 \times 14 = 2{,}016$.

 Find more at
ViewMath.com/FL-Grade6

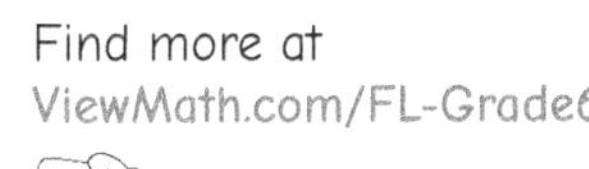

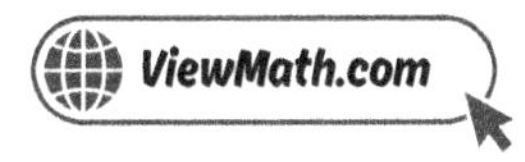

8.   Multiples of 3: $3, 6, 9, \ldots$ Multiples of 9: $9, 18, \ldots$ The smallest number in both lists is 9. When one number is a multiple of the other, the LCM is the larger number.

9.   The opposite of -4 is 4. Both -4 and 4 are 4 units from zero, but on opposite sides of the number line.

10.   The x-coordinate -3 tells you to move 3 units to the left, and the y-coordinate 4 tells you to move 4 units up. Choice D swaps the coordinates.

11.   Work left to right: $(-2) \times 5 = -10$. Then $(-10) \times (-3) = 30$. Then $30 \div (-6) = -5$. Alternatively, count negatives in (-2), (-3), (-6): 3 negatives (odd) $\rightarrow$ negative. $2 \times 5 \times 3 \div 6 = 5$, so the answer is -5.

12.   In Step 2, the student added $2 + 5$ before multiplying. By PEMDAS, exponents come first ($3^2 = 9$), then multiplication ($5 \times 9 = 45$), then addition ($2 + 45 = 47$).

13.   First find the sum $a + b$, then divide by 2: $\frac{a+b}{2}$.

14. $3(p + 4)$ means $3 \times (p + 4)$. The group $(p + 4)$ is one of the two factors being multiplied.

15. $15 \div 3 + 8 = 5 + 8 = 13$.

16.   Subtract 3.5: $x = 10 - 3.5 = 6.5$.

17.   "Less than 50" means strictly under 50: $w < 50$.

18.   Any number greater than -1 is a solution. -1 itself is not, since $>$ does not include the boundary.

19.   The independent variable (number of cars) goes on the x-axis.

20.   Each triangle: $\frac{1}{2} \times 6 \times 8 = 24$ in^2. Two triangles: $24 \times 2 = 48$ in^2.

Find more at
ViewMath.com/FL-Grade6

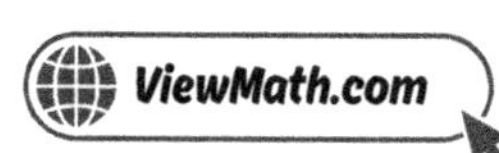

21. $V = l \times w \times h$. If h becomes $2h$, then $V_{new} = l \times w \times 2h = 2(lwh)$, so the volume doubles.

22. $|8 - (-4)| = |12| = 12$ units.

23. $8 \times 5 \times 2 = 80$ is the volume. The actual surface area is $2(40) + 2(16) + 2(10) = 132 \ cm^2$.

24. Exercise time varies from student to student. The other questions each have one fixed answer.

25. Distances from 24: $4, 2, 0, 2, 4$. $MAD = (4 + 2 + 0 + 2 + 4) \div 5 = 12 \div 5 = 2.4$.

26. Favorite color is categorical, not numerical. A frequency table or bar graph handles categories; histograms and dot plots need numerical data.

27. Different data sets can have the same five-number summary. The five-number summary only captures five key values, not every data point.

28. Week 2 median (9,500) is higher. Week 2 IQR (1,200) is smaller, meaning more consistent daily step counts.

29. A fair coin has 2 equally likely outcomes (heads or tails). $P(heads) = \dfrac{1}{2}$.

30. In a stem-and-leaf plot the stem is formed by all digits except the last digit, and the leaf is the last digit.

✔ Practice Test 7 — Answer Key

1. 6

2. B

3. $(2, 5)$ and $(8, 20)$ (or any equivalent pair)

4. B

5. Advantage: You can buy something now even if you do not have enough cash. Disadvantage: If you do not pay the balar

Find more at
ViewMath.com/FL-Grade6

 C    216    C    B    B    −$40    D    C    B

 45    A    B    B    C    7 cm    B    B    142 m^2

 B    20    C

 *Store B has higher typical sales (median ≈ $275 vs. $250). Store A has more predictable sales (IQR ≈ $100 vs. $*

 *B   29 B   30 C*

💡 Time to Learn! 💡

*Review the explanations below, **especially for the questions you missed**.*

Understanding why each answer is correct builds stronger problem-solving skills.

Tip: *Circle any questions you got wrong, then read their explanation carefully.*

📖 Practice Test 7 — Detailed Explanations

1. $21 \div 7 = 3$, *so multiply both by 3: iced tea = $2 \times 3 = 6$.*

2. $2 \times 4 = 8$ *and* $3 \times 4 = 12$, *so* $8 : 12$ *is equivalent to* $2 : 3$.

3. *Ratio* $= 4 : 10 = 2 : 5$. *Half gives* $(2, 5)$; *double gives* $(8, 20)$.

4. 70% *of* $30 = 0.70 \times 30 = 21$ *students.*

5. *Credit cards allow purchases without immediate cash but can lead to debt through interest charges if balances are not paid in full.*

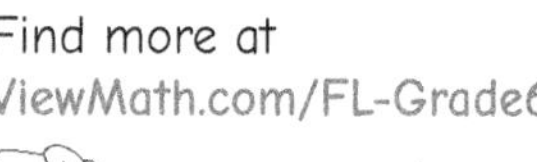

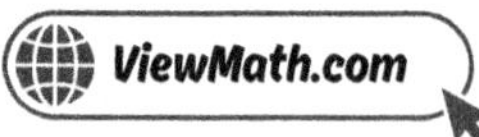

Keep, Change, Flip: $\frac{1}{2} \times \frac{4}{1} = \frac{4}{2} = 2$.

$90 \div 42 = 2$ R6; bring down 7 $\to$ $67 \div 42 = 1$ R25; bring down 2 $\to$ $252 \div 42 = 6$. Answer: 216. Check: $216 \times 42 = 9{,}072$.

Factors of 60: $1, 2, 3, 4, 5, 6, 10, 12, 15, 20, 30, 60$. Factors of 84: $1, 2, 3, 4, 6, 7, 12, 14, 21, 28, 42, 84$. Common factors: $1, 2, 3, 4, 6, 12$. The greatest is 12.

The opposite of a number is the same distance from zero on the other side of the number line. The opposite of 9 is -9 because both are 9 units from zero.

When the y-coordinate is 0, the point lies on the x-axis. The point $(-6, 0)$ is 6 units to the left of the origin on the x-axis. A common mistake is thinking a negative x-value means the point is on the y-axis.

Each day is $-\$8$. Over 5 days: $5 \times (-8) = -40$. The stock lost \$40 total.

The base 5 is multiplied 3 times: 5^3.

5 tickets at d dollars each costs $5 \times d = 5d$.

The constants are 3 and 8 — terms with no variable. $5x$ and $2y$ are variable terms.

$8(6) - 3 = 48 - 3 = 45$.

Divide both sides by 5: $n = 45 \div 5 = 9$.

"Greater than 7" uses the $>$ symbol: $x > 7$.

$x > 2$ does NOT include 2, so the circle should be open, not closed.

Find more at
ViewMath.com/FL-Grade6

19. *Time (hours) is independent — it passes regardless. Battery level depends on time.*

20. $84 = \frac{1}{2}(10 + 14)h = \frac{1}{2}(24)h = 12h.$ *So* $h = 84 \div 12 = 7$ *cm.*

21. $V = 3 \times 2 \times 2 = 12$ *ft*^{3}.

22. *For a horizontal side (same y), subtract the x-coordinates and take the absolute value.*

23. $SA = 2(7)(5) + 2(7)(3) + 2(5)(3) = 70 + 42 + 30 = 142$ *m*^{2}.

24. *The dot plot shows data ranging from 0 to 5, with different students giving different answers. This variability confirms the question is statistical.*

25. *Median* $= 115$. *Q1: median of* $100, 105, 110 = 105$. *Q3: median of* $120, 125, 130 = 125$. *IQR* $= 125 - 105 = 20$.

26. *Every value is 5, so 5 is the value that appears most often. The mode is 5.*

27. *Store A: median* $\approx \$250$, *IQR* $\approx 300 - 200 = \$100$. *Store B: median* $\approx \$275$, *IQR* $\approx 350 - 225 = \$125$. *Store B sells more on a typical day, but Store A's sales are more consistent (smaller IQR and range).*

28. *The median describes a typical value (center) and the IQR describes how spread out the middle 50% of values are (spread). Together they summarize the data well.*

29. $\frac{2}{5} = 0.4$ *and* $\frac{3}{4} = 0.75$. *Since* $0.75 > 0.4$, *Event B is more likely.*

30. *There are 9 values, so the median is the 5th value. In order:* $20, 24, 31, 35, 3?, 42, 46, 48, 53$. *The 5th value must equal* 37, *so the missing leaf is 7.*

☑ Practice Test 8 — Answer Key

1 B   **2** Part A: $2 : 1$;   Part B: 4 cups of sugar

3 False. Equivalent ratios always form a straight line through the origin.   **4** C

5 Interest: $96. Total: $896.   **6** D   **7** A   **8** B   **9** B

10 Part A: Quadrant II;   Part B: $(8, 1)$, Quadrant I;   Part C: $(-8, -1)$, Quadrant III   **11** C   **12** B

13 $12h + 20$ **14** $A = 4$, $B = (x + 3)$ **15** 25   **16** $a = 7$

17 $n = 4$ is NOT a solution to $n > 4$ (since 4 is not greater than 4). $n = 4$ IS a solution to $n \geq 4$ (since 4 equals 4).

18 C   **19** $w = 45 - 3t$; empty when $t = 15$ minutes   **20** B   **21** $1\ ft^3$ **22** B   **23** B

24 Answers vary. Example: "How many glasses of water do you drink each day?"   **25** 6   **26** B

27 50%   **28** C   **29** 20%   **30** B

💡 Time to Learn! 💡

Review the explanations below, **especially for the questions you missed**.

Understanding why each answer is correct builds stronger problem-solving skills.

Tip: Circle any questions you got wrong, then read their explanation carefully.

📖 Practice Test 8 — Detailed Explanations

1 Dogs have 4 parts. Each part $= 2$ animals. Dogs $= 4 \times 2 = 8$.

Find more at
ViewMath.com/FL-Grade6

(2) Part A: From the graph, $(2, 1)$ shows 2 cups of flour for every 1 cup of sugar. Part B: Following the pattern, $(8, 4)$, so 4 cups of sugar.

(3) When you multiply both parts of a ratio by the same number, the points increase at a constant rate, which produces a straight line through $(0, 0)$.

(4) $100\% = \dfrac{100}{100} = 1$ whole.

(5) $I = 800 \times 0.06 \times 2 = \$96$. Total $= \$800 + \$96 = \$896$.

(6) The student flipped the first fraction ($\dfrac{4}{5} \to \dfrac{5}{4}$) instead of the second ($\dfrac{2}{3} \to \dfrac{3}{2}$). The correct answer is $\dfrac{4}{5} \times \dfrac{3}{2} = \dfrac{12}{10} = \dfrac{6}{5}$.

(7) To check a division, multiply the quotient by the divisor. If $4{,}752 \div 12 = 396$, then 396×12 should equal $4{,}752$.

(8) $36 \div 8 = 4.5$, so 8 does not divide evenly into 36. It is a factor of 24 but not of 36, so it is not a common factor. The other choices (4, 6, 12) all divide evenly into both 24 and 36.

(9) "How much someone owes" is measured by absolute value. $|-35| = 35$ and $|-20| = 20$. Since $35 > 20$, Riley owes more. A more negative balance means a larger debt.

(10) Part A: $(-8, 1)$ has signs $(-, +)$, which is Quadrant II. Part B: Reflecting across the y-axis flips the x-sign: $(-8, 1) \to (8, 1)$. Signs $(+, +)$ is Quadrant I. Part C: Reflecting across the x-axis flips the y-sign: $(-8, 1) \to (-8, -1)$. Signs $(-, -)$ is Quadrant III.

(11) The number line shows 3 jumps of -4 starting at 0 and ending at -12. This represents $3 \times (-4) = -12$.

(12) Parentheses: $5 - 1 = 4$. Exponent: $4^2 = 16$. Add: $16 + 7 = 23$.

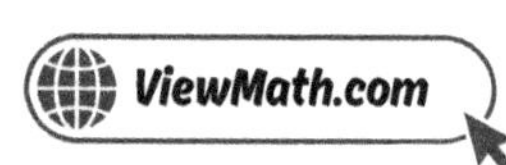

Hourly pay is $12h$. Plus the bonus: $12h + 20$.

$4(x + 3) = 4 \times (x + 3)$. The two factors are 4 and $(x + 3)$.

$4^2 + 2(4) + 1 = 16 + 8 + 1 = 25$.

$a = 105 \div 15 = 7$.

$>$ does not include equality. $\geq$ includes equality.

$-3 \geq -3$ is true (they are equal). The other values are less than -3.

Set $w = 0$: $0 = 45 - 3t$, so $3t = 45$, $t = 15$ minutes.

The student used the slant side instead of the height. Area $= 8 \times 5 = 40$ square units.

$V = \frac{1}{3} \times \frac{1}{2} \times 6 = \frac{6}{6} = 1$ ft^3.

Same y-coordinate, so it is a horizontal distance. $|6 - (-4)| = |10| = 10$ units.

A net is a flat pattern that, when folded, forms a 3D shape. Each section of the net becomes a face.

Any question about food where different classmates would give different answers is acceptable.

Distances from 60: $10, 5, 0, 5, 10$. $MAD = (10 + 5 + 0 + 5 + 10) \div 5 = 30 \div 5 = 6$.

Favorite fruit is categorical data (not numerical intervals). A frequency table counts each category and is the natural first step.

Find more at
ViewMath.com/FL-Grade6

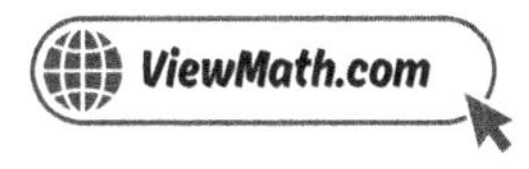

27 *Q1 to Q3 is the box, which always contains 50% of the data.*

28 *The median and IQR are resistant to outliers. The mean and range can be greatly affected by extreme values.*

29 *Total marbles:* $4 + 6 + 10 = 20$. $P(green) = \dfrac{4}{20} = \dfrac{1}{5} = 0.20 = 20\%$.

30 *Values greater than 50:* $51, 53, 56, 62$. *That is 4 values.*

✅ Practice Test 9 — Answer Key

1 C   2 C   3 A   4 C   5 C   6 A   7 B

8 Part A: $1, 2, 3, 4, 6, 8, 12, 24$;   Part B: 24;   Part C: 24 boxes, each with 2 cookies and 3 brownies   9 C

10 Part A: $A = (2, 5)$, $B = (5, 1)$, $C = (2, 1)$;   Part B: $A' = (-2, 5)$, $B' = (-5, 1)$, $C' = (-2, 1)$;   Part C: Quadrant I

11 D   12 C   13 B   14 B   15 C   16 C   17 C   18 B   19 C   20 C

21 B   22 C   23 B   24 C   25 B   26 B   27 A   28 B   29 B   30 28

💡 Time to Learn! 💡

*Review the explanations below, **especially for the questions you missed**.*

Understanding why each answer is correct builds stronger problem-solving skills.

Tip: *Circle any questions you got wrong, then read their explanation carefully.*

📖 Practice Test 9 — Detailed Explanations

Find more at
ViewMath.com/FL-Grade6

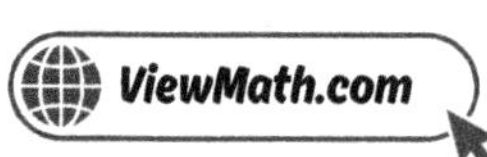

1. *Total parts $= 1 + 3 + 2 = 6$. Each part $= 18 \div 6 = 3$. Blue $= 3 \times 3 = 9$.*

2. *$1 \times 3 = 3$ red, so $4 \times 3 = 12$ blue. Multiply both parts by 3.*

3. *Line A rises 4 for every 1 it moves right; Line B rises 3. A higher rise means steeper.*

4. *88 out of $100 = 88\%$. When the total is 100, the number correct IS the percent.*

5. *He saves $\$80 - \$50 = \$30$. Percent saved $= \frac{30}{80} \times 100 = 37.5\%$.*

6. *$\frac{3}{4} \div \frac{1}{8} = \frac{3}{4} \times \frac{8}{1} = \frac{24}{4} = 6$ pieces.*

7. *$26 \div 16 = 1$ R10; bring down $8 \to 108 \div 16 = 6$ R12; bring down $8 \to 128 \div 16 = 8$. Answer: 168. Check: $168 \times 16 = 2{,}688$.*

8. *Part A: Numbers appearing in both rows: $1, 2, 3, 4, 6, 8, 12, 24$. Part B: The greatest of these common factors is 24. Part C: $48 \div 24 = 2$ cookies per box and $72 \div 24 = 3$ brownies per box.*

9. *Absolute value measures distance from zero, and distance is never negative. So absolute value is always ≥ 0. $|-5| = 5$ (not -5), and $|0| = 0$.*

10. *Part A: Reading from the grid, $A = (2, 5)$, $B = (5, 1)$, $C = (2, 1)$. Part B: Reflecting across the y-axis changes the sign of each x-coordinate: $A' = (-2, 5)$, $B' = (-5, 1)$, $C' = (-2, 1)$. Part C: All reflected vertices have negative x-coordinates and positive y-coordinates, which is the sign convention $(-, +)$ for Quadrant II.*

11. *Count the negative factors: there are 3 (odd number), so the product is negative. $2 \times 3 \times 4 = 24$, so the answer is -24.*

12. *$4^3 = 4 \times 4 \times 4 = 64$.*

13. *"Triple b" is $3b$. "Then subtract 4" gives $3b - 4$.*

14. *When no number is written in front of a variable, the coefficient is 1: $m = 1 \cdot m$.*

15. *$3^2 + 4^2 = 9 + 16 = 25$.*

16. *Multiply by 6: $t = 5 \times 6 = 30$.*

17. *$5 > 5$ is false. 5 equals 5, but is not greater than 5. It would be a solution to $x \geq 5$.*

18. *Open circle means -2 is NOT included ($<$ or $>$). Shading left means less than: $x < -2$.*

19. *$c = 7(6) = 42$ dollars.*

20. *Area $= 25 \times 14 = 350$ m^2. Bags needed: $350 \div 50 = 7$ bags.*

21. *Box A volume $= 6 \times 4 \times 5 = 120$. Box B: $120 = 10 \times 3 \times h = 30h$, so $h = 4$.*

22. *Same x-coordinate, so it is a vertical distance. $|5 - (-3)| = |8| = 8$ units.*

23. *$6s^2 = 96$, so $s^2 = 16$, giving $s = 4$ cm.*

24. *$12 \times 9 = 108$ is a single fixed answer. It does not vary from person to person.*

25. *Data Set A: range $= 9 - 5 = 4$. Data Set B: range $= 10 - 4 = 6$. Data Set B has the larger range.*

26. *The size of the data set and whether it is numerical or categorical determine the best display. Categorical ⯎ frequency table/bar graph. Small numerical ⯎ dot plot. Large numerical ⯎ histogram.*

27. Class A's box (IQR) is narrower than Class B's. A narrower box means the middle 50% of scores are closer together — more consistent.

28. The range uses only the maximum and minimum. Changing any extreme value changes the range. However, the IQR only uses Q_1 and Q_3 and can stay the same even if extremes change. Still, the range is most directly influenced by outliers since it depends on the two most extreme values.

29. $P(cherry) = \dfrac{4}{10}$ and $P(grape) = \dfrac{4}{10}$. The probabilities are equal, so these events are equally likely.

30. The values are $20, 24, 28, 32, 36$. Sum $= 20 + 24 + 28 + 32 + 36 = 140$. Mean $= 140 \div 5 = 28$.

✅ Practice Test 10 — Answer Key

1. C **2.** B **3.** C **4.** A **5.** B **6.** $\dfrac{2}{3} \div \dfrac{1}{6} = 4$ **7.** 252 **8.** 28 seconds

9. C **10.** D

11. Part A: $(-3) \times (-1) = 3$, $(-3) \times (-2) = 6$, $(-3) \times (-3) = 9$. Part B: Each time the second factor decreases by 1, the pro

12. B **13.** $4x + 5$ **14.** D **15.** B **16.** $m = 13$ **17.** C **18.** $x \le 4$ **19.** C **20.** B

21. B **22.** C **23.** 247.2 cm^2 **24.** C **25.** C **26.** 11 **27.** B **28.** C **29.** C

30. B

Find more at
ViewMath.com/FL-Grade6

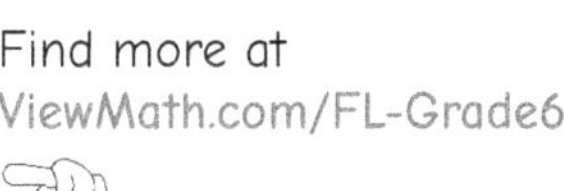

💡 Time to Learn! 💡

Review the explanations below, **especially for the questions you missed**.

Understanding why each answer is correct builds stronger problem-solving skills.

Tip: Circle any questions you got wrong, then read their explanation carefully.

📖 Practice Test 10 — Detailed Explanations

1. Apple has 3 parts, orange has 5 parts. Total parts = 8. Total ounces = $8 \times 4 = 32$.

2. The ratio is $2 : 50$. For 6 bags: $50 \times 3 = 150$ sq ft.

3. At \$2 per bottle, 4 bottles cost \$8, not \$10. The point $(4, 10)$ does not fit the ratio.

4. Convert all to decimals: $30\% = 0.30$, $\frac{1}{3} \approx 0.333$, 0.35. Order: $0.30 < 0.333 < 0.35$.

5. Credit interest owed: $\$150 \times 0.10 = \$15$. Savings interest earned: $\$150 \times 0.02 = \$3$. Difference: $\$15 - \$3 = \$12$.

6. $\frac{2}{3} \div \frac{1}{6} = \frac{2}{3} \times \frac{6}{1} = \frac{12}{3} = 4$. The shaded part of Bar B fits into the shaded part of Bar A exactly 4 times.

7. $37 \div 15 = 2$ R7; bring down $8 \to 78 \div 15 = 5$ R3; bring down $0 \to 30 \div 15 = 2$. Answer: 252. Check: $252 \times 15 = 3{,}780$.

8. Find the LCM of 4 and 7. Multiples of 4: $4, 8, 12, 16, 20, 24, 28, \ldots$ Multiples of 7: $7, 14, 21, 28, \ldots$ LCM $= 28$. They blink together again after 28 seconds.

9. A number and its opposite are the same distance from zero on different sides, so they cancel each other out. For example, $6 + (-6) = 0$ and $-9 + 9 = 0$.

10. Quadrant IV has sign convention $(+, -)$. Since $2 > 0$ and $-5 < 0$, the point $(2, -5)$ is in Quadrant IV. A common error is confusing Quadrant IV with Quadrant II which has signs $(-, +)$.

11. The products are $-9, -6, -3, 0, \ldots$ increasing by 3 each time. Continuing: $0 + 3 = 3$, $3 + 3 = 6$, $6 + 3 = 9$. The pattern demonstrates that multiplying two negatives yields a positive.

12. 1 raised to any power equals 1: $1^{10} = 1$.

13. There are 4 triangles (each worth x) and 5 unit squares. The expression is $4x + 5$.

14. $x = 1 \cdot x$, so the coefficient is 1. The expression has 2 terms: x and 5.

15. $10 - 2(4 - 1) = 10 - 2(3) = 10 - 6 = 4$.

16. Subtract 17: $m = 30 - 17 = 13$.

17. "More than \$25" means greater than 25: $d > 25$.

18. Closed circle at 4 (included) and shading to the left (less than): $x \leq 4$.

19. Each dog has 4 legs: $L = 4d$.

20. $A = \frac{1}{2}(5 + 13)(8) = \frac{1}{2}(18)(8) = 72 \ ft^2$.

21. $V = 5 \times 3 \times 4 = 60 \ cm^3$.

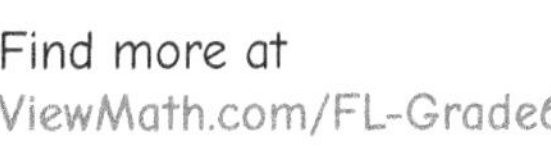

Find more at
ViewMath.com/FL-Grade6

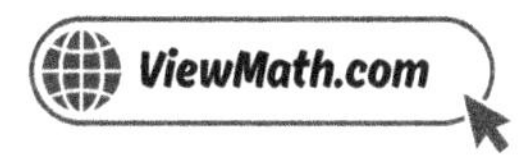

22 *The base goes from $(0,0)$ to $(8,0)$. Distance $= |8-0| = 8$ units.*

23 *2 triangles: $2 \times \frac{1}{2}(6)(5.2) = 31.2$. 3 rectangles: $3 \times 6 \times 12 = 216$. Total: $31.2 + 216 = 247.2$ cm^2.*

24 *If you ask 10 people how long they spend eating lunch, the answers vary. Following the flowchart, varying answers lead to "Statistical."*

25 *Adding 5 to every value shifts the max and min by the same amount, so the difference (range) stays unchanged.*

26 *0–14: 3 people. 15–29: 8 people. Total < 30 min: $3 + 8 = 11$.*

27 *The box stretches from $Q1$ to $Q3$, containing the middle 50% of the data.*

28 *A: $IQR = 30 - 15 = 15$. B: $IQR = 35 - 20 = 15$. Both have $IQR = 15$.*

29 *Total balls: $5 + 3 + 2 = 10$. $P(yellow) = \dfrac{2}{10}$. $P(not\ yellow) = 1 - \dfrac{2}{10} = \dfrac{8}{10}$.*

30 *Stem 5 has 5 leaves, which is more than any other stem.*

Well done checking your answers!

Keep practicing to strengthen your skills

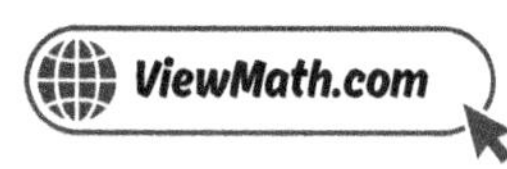

Author's Final Note

I hope you enjoyed this book as much as I enjoyed writing it. Whether you are a student working through the material, a parent supporting your child's learning, or a teacher guiding your class, I have tried to make this book as clear and engaging as possible. I hope I have succeeded. If you have any suggestions for improvement, please let me know. I would love to hear from you.

The accuracy of calculations is very important to me. We have done our best, but I also expect that I have made some minor errors. Constant improvement is the name of the game. If you find any errors, please let me know. I will fix them in the next edition.

For students: Your learning journey does not end here. I have written a series of books to help you learn math. Make sure you browse through them. I especially recommend workbooks and practice tests to help you prepare for your exams.

For parents: Thank you for investing in your child's education. I encourage you to explore the companion resources available online to help support your child outside the classroom.

For teachers: Thank you for the invaluable work you do every day. I hope this book serves as a useful resource in your classroom. Feel free to reach out if you have suggestions or would like to discuss how best to use this book with your students.

I also enjoy reading your reviews. If you have a moment, please leave a review on where you found this book. It will help others find this book. If you have any questions or comments, please feel free to contact me at drNazari@ViewMath.com.

And one last thing: Remember to use online resources for additional help. I recommend using the resources on `https://ViewMath.com` You can find video lessons, practice problems, and more. You can also use the online companion for this book to track your progress and access additional resources.

Wishing all students the best in their studies, parents every success in supporting their children, and teachers continued inspiration in their classrooms!

Dr. A. Nazari

 # Great Job! Keep Learning with ViewMath!

*Keep up the great work! Visit **viewmath.com/FL-Grade6** for free lessons, quizzes, and more.*

Study Guide

Workbook

Step-by-Step

3 Practice Tests

5 Practice Tests

7 Practice Tests

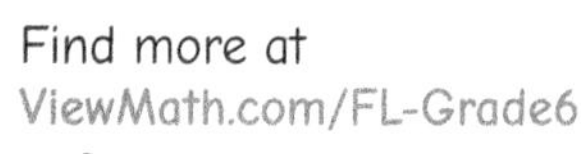

Find more at
ViewMath.com/FL-Grade6

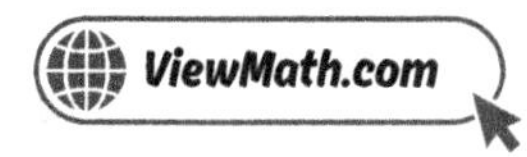